Neuronal Cel
A Practical Approach

Edited by
JOHN N. WOOD

*Sandoz Institute for Medical Research,
London*

at
OXFORD UNIVERSITY PRESS
Oxford New York Tokyo

Oxford University Press, Walton Street, Oxford OX2 6DP

Oxford New York Toronto
Delhi Bombay Calcutta Madras Karachi
Petaling Jaya Singapore Hong Kong Tokyo
Nairobi Dar es Salaam Cape Town
Melbourne Auckland
and associated companies in
Berlin Ibadan

Oxford is a trade mark of Oxford University Press

A Practical Approach ⬡ is a registered trade mark
of the Chancellor, Masters, and Scholars of the University of Oxford
trading as Oxford University Press

Published in the United States
by Oxford University Press, New York

© Oxford University Press, 1992

All rights reserved. No part of this publication may be
reproduced, stored in a retrieval system, or transmitted, in any
form or by any means, without the prior permission in writing of Oxford
University Press. Within the UK, exceptions are allowed in respect of any
fair dealing for the purpose of research or private study, or criticism or
review, as permitted under the Copyright, Designs and Patents Act, 1988, or
in the case of reprographic reproduction in accordance with the terms of
licences issued by the Copyright Licensing Agency. Enquiries concerning
reproduction outside those terms and in other countries should be sent to
the Rights Department, Oxford University Press, at the address above.

This book is sold subject to the condition that it shall not,
by way of trade or otherwise, be lent, re-sold, hired out, or otherwise
circulated without the publisher's prior consent in any form of binding
or cover other than that in which it is published and without a similar
condition including this condition being imposed
on the subsequent purchaser.

Users of books in the Practical Approach Series are advised that prudent
laboratory safety procedures should be followed at all times. Oxford
University Press makes no representation, express or implied, in respect of
the accuracy of the material set forth in books in this series and cannot
accept any legal responsibility or liability for any errors or omissions
that may be made.

A catalogue record for this book is available from the British Library

Library of Congress Cataloging in Publication Data
Neuronal cell lines : a practical approach / edited by John N. Wood.
(Practical approach series)
Includes bibliographical references and index.
1. Nerve tissue—Cultures and culture media. I. Wood, John N.
II. Series
[DNLM: 1. Cell Line. 2. Neurons—physiology. WL 102.5 C3925]
QP361.C39   1992   596'.0188—dc20   92-1439
ISBN 0-19-963346-0 (hbk.)
ISBN 0-19-963345-2 (pbk.)

Typeset by Footnote Graphics, Warminster, Wilts
Printed and bound by Information Press Ltd, Oxford, England

Preface

The development and function of the vertebrate nervous system is now one of the most rapidly expanding areas of biological research. Thanks to technical advances in molecular genetics, electrophysiology, and cell biology, it is now possible to examine the functional activity of neuronal subsets which were previously identified solely by anatomical or morphological criteria. *In vitro* studies of isolated tissue, primary neuronal culture, and neuronal cell lines all have a role to play. Tissue slices and roller cultures of isolated brain areas allow the *in vitro* examination of functionally intact neuronal circuits. Primary culture of dissociated neuronal tissue allows an analysis of the factors responsible for neuronal survival and function, and enables the cellular physiology of isolated neurons and glia to be explored. Finally, cell lines that retain some of the properties of neural cells provide an invaluable tool for a variety of biochemical and mechanistic studies which are difficult or impossible to carry out with small numbers of heterogenous neurons in culture, and the present volume addresses these areas. The companion volume in *The Practical Approach* series entitled *Cellular Neurobiology* covers related topics, whilst concentrating on the use of primary neuronal cultures.

A familiarity with tissue culture is assumed here, and issues such as mycoplasma screening, the importance of passing cells at appropriate densities and the significance of passage number are not covered. A companion volume (*Animal Cell Culture*) provides important information with respect to these fundamental, yet often ignored, aspects of tissue culture.

Neuronal cell lines are important and useful for one principal reason: they are easy to grow and provide a homogeneous population of cells. The identification and characterization of molecules, ranging from growth factors to ion channels, and the regulatory events ranging from transcriptional activation to second-messenger actions on ion fluxes, can all be rapidly addressed in such lines. Indeed, the investigation of neuronal cell lines may provide such a wealth of information that the temptation exists to regard such model systems as legitimate research targets in their own right. This is not the aim of the present publication. Cell lines provide useful information that must subsequently be validated *in vivo*. We hope that the techniques described in this book will prove useful for the exploitation of cell lines by neurobiologists in a variety of disciplines, and will provide insights into the technology of cell line generation and the possibilities afforded by transgenic technology.

London J.N.W.
December 1991

Contents

List of contributors   xv

Abbreviations   xvii

1. Neuronal hybrid cell lines: generation, characterization, and utility   1
Bruce H. Wainer and Alfred Heller

   1. Introduction   1
   2. Methodology   4
      Dissection and dissociation of primary mouse brain cells   4
      Neuroblastoma fusion partner   6
      Cell fusion   7
      General characterization of hybrid cell lines   9
   3. Septal and hippocampal hybrid cell lines   11
      Septal cell lines   11
      Hippocampal cell lines   15
   4. Mesencephalic and striatal cell lines   16
      Mesencephalic cell lines   17
      Striatal cell lines   20
   5. Conclusions and future directions   21
   Acknowledgements   24
   References   24

2. Generation of neural cell lines by transfer of viral oncogenes   27
Pierre Rouget, Marc Le Bert, Isabelle Borde, and Claudine Evrard

   1. Introduction   27
   2. Preparation of cell cultures for transfection   28
      Transient expression vs. integration   28
      Choice of developmental stage   29
      Cell culture conditions before transfection or transduction   29
   3. Transformation with oncogenic DNA viruses   32
   4. Transfection of cells with oncogenes   33
      Oncogenes and vectors for transfection-mediated immortalization   33
      Calcium phosphate precipitation   35

Electroporation	36
Lipofection	37

5. Transduction of oncogenes with recombinant retroviral vectors — 38
- Recombinant retroviral vectors — 38
- Trans-complementation between the vector and helper sequences — 39
- Typical recombinant retroviral vectors carrying an oncogene — 39
- Packaging cell lines — 41
- Production of recombinant retrovirus stock — 42
- Determining the recombinant retrovirus titre — 42
- Retroviral vector-mediated transduction — 43

6. Selection and growth properties of cell lines — 44
- Selection and establishment of cell lines — 44
- Growth properties of cells: immortalization vs. transformation — 45
- Status of the transferred oncogene and clonality of cell lines — 46
- Expression of the oncogene — 46
- Precautions for avoiding secondary transformation events — 47

7. Characterization of cell lines: differentiation and functional properties — 47
- General considerations on the differentiation of permanent cell lines — 47
- Induction of differentiation — 48
- Main cell-type specific differentiation markers — 48
- Other markers and functional properties — 50

8. Future prospects for generating cell lines from transgenic mice — 50

References — 52

3. Serum-free media for neuronal cell culture 55
Michael Butler

1. Introduction — 55

2. The development of serum-free media — 56
- The analytical approach — 56
- Ham's approach — 58
- Sato's approach — 60

3. Defined media supplements — 61
- Hormones — 61
- Growth factors — 63
- Attachment factors — 65
- Carrier proteins — 66
- Lipids — 66
- Trace elements — 66

4. Serum-free formulations for specific neural cell types — 67

Neuronal cell lines	67
Primary neurons	70
Glial cell lines	71
Primary glial cells	71
5. Undefined low protein serum substitutes	72
6. Commercially available serum substitutes	72
7. Conclusion	72
References	73

4. Embryonal carcinoma cells and embryonic stem cells as models for neuronal development and function 77
James W. McCarrick and Peter W. Andrews

1. Introduction	77
2. Teratocarcinomas as tools in embryology	77
Origins of teratocarcinomas and their relationship to the early embryo	77
Genetic manipulation using ES cells	79
3. Cell lines of relevance to neurobiology: their maintenance and differentiation	79
Introduction	79
EC cell lines	80
ES cell lines	89
4. Stem cell markers and monitoring differentiation	92
5. Production of mutant mice by homologous recombination in ES cells	94
General considerations	94
Design of gene targeting constructs	95
Introduction of DNA into ES cells	97
Selection	100
Isolation and characterization of clones	101
Production of chimeras	102
Acknowledgements	102
References	102

5. Correlative electrophysiological and biochemical studies in neuronal cell lines 105
Philip M. Dunn, Paul R. Coote, and John N. Wood

1. Introduction	105
Introduction to electrophysiological principles	105
2. Modulation of intracellular second-messenger levels	107

Measurement of intracellular second messengers	111
Biochemical measurement of ion fluxes	115

3. Electrophysiological effects of neuroactive ligands — 117
Electrophysiological actions of neurotransmitters, neuromodulators, and drugs — 117
Electrophysiological characterization of drug responses — 118

4. Electrophysiological recording techniques — 119
Equipment and practical considerations — 120
Sharp microelectrode recording — 121
Determination of current–voltage relationships and reversal potential — 123
'Patch-clamp' recording — 124
Methods of drug application — 126

5. Correlating changes in second-messenger levels with altered ion fluxes — 130

References — 131

6. Analysis of protein phosphorylation in cell lines   133
James R. Woodgett

1. Introduction — 133

2. Metabolic labelling of cells in culture — 133
Incorporation of ^{32}P — 133
Labelling conditions — 135
Biosynthetic labelling of proteins — 137

3. Analysis of phosphoproteins — 138
Two-dimensional gel analysis — 138
Detection of phosphotyrosine-containing proteins using specific antibodies — 141
Immunopurification — 141
Enrichment for phosphotyrosine — 144
Phosphoamino acid determination — 145
Phosphopeptide mapping — 147

4. Identification of phosphorylation sites: strategies — 151
Secondary proteolytic digestion — 152
Manual Edman degradation — 152
Prediction of migration of phosphopeptides upon electrophoresis and chromatography — 153
Use of synthetic peptides — 153
Strategies for identification of relevant protein kinases and phosphatases — 156
Assay of protein kinases blotted on to membranes — 157
A final word about protein phosphatases — 158

References — 159

Contents

7. Isolation and characterization of transfected cell lines expressing neurotransmitter receptors using a calcium dye and flow cytometry 161

Ira Schieren and Amy B. MacDermott

1. Introduction	161
2. Mechanisms of $[Ca^{2+}]_i$ increase	162
3. Calcium indicator	163
4. Flow cytometry	166
Hardware	166
Data display and analysis	168
Calibration	170
5. Application	174
Identification and selection of cells with functional receptors	174
Physiology	175
Pharmacology	177
6. Conclusion and summary	177
References	179

8. Regulation of gene expression in neuronal cell lines 181

Karen A. Lillycrop, Carolyn L. Dent, and David S. Latchman

1. Introduction	181
2. Gene expression	181
Detection and quantitation of RNA	182
Measurement of transcription rates	194
3. Promoter activity	197
Transfection	197
Assay of promoter activity	197
4. Identification of DNA-binding transcription factors	203
Preparation of cellular extracts	203
DNA mobility shift assay	204
DNase I footprinting assay	206
Methylation interference assays	209
South-western blotting	211
5. Conclusion	213
References	214

9. Cell lines in developmental neurobiology: assays of adhesion and neurite outgrowth 217
John A. Pizzey

1. Introduction	217
2. Intercellular adhesion	218
Basic principles	218
3. Intercellular adhesion assays	224
Basic principles	224
Measurements of specific adhesion	232
4. Neurite outgrowth	237
Basic principles	237
Neurite outgrowth assays	238
5. Future prospects	244
References	245

Appendices
A1 A catalogue of neuronal properties expressed by cell lines 249
Iain F. James and John N. Wood

Introduction	249
Table 1: Cell lines that express neuropeptides	249
Table 2: Cell lines that express peptide receptors	250
Table 3: Cell lines expressing neurotransmitter receptors	252
Table 4: Cell lines expressing voltage-activated ion channels	253
Table 5: Sources of cell lines	254
References	254

A2 Suppliers of specialist items 261

Index 263

Contributors

PETER W. ANDREWS
The Wistar Institute, 3601 Spruce Street, Philadelphia, PA 19104, USA.

MARC LE BERT
Laboratoire de Biochemie Cellulaire, Collège de France, 11 Place Marcelin Berthelot, 75231 Paris cedex 05, France.

ISABELLE BORDE
Laboratoire de Biochemie Cellulaire, Collège de France, 11 Place Marcelin Berthelot, 75231 Paris cedex 05, France.

MICHAEL BUTLER
Department of Microbiology, University of Manitoba, Winnipeg R3T 2N2, Canada.

PAUL R. COOTE
Department of Neuroimmunology, Sandoz Institute for Medical Research, 5 Gower Place, London WC1E 6BN, UK.

CAROLYN L. DENT
Medical Molecular Biology Unit, Department of Biochemistry, University College and Middlesex School of Medicine, Windeyer Building, Cleveland Street, London W1P 6DB, UK.

PHILIP M. DUNN
Department of Pharmacology, University College, Gower Street, London WC1, UK.

CLAUDINE EVRARD
Laboratoire de Biochemie Cellulaire, Collège de France, 11 Place Marcelin Berthelot, 75231 Paris cedex 05, France.

ALFRED HELLER
Department of Pharmacological and Physiological Sciences, University of Chicago, 947 East 58th Street, Chicago, IL 60637, USA.

IAIN F. JAMES
Department of Biochemistry, Sandoz Institute for Medical Research, 5 Gower Place, London WC1E 6BN, UK.

Contributors

DAVID S. LATCHMAN
Medical Molecular Biology Unit, Department of Biochemistry, University College and Middlesex School of Medicine, Windeyer Building, Cleveland Street, London W1P 6DB, UK.

KAREN A. LILLYCROP
Medical Molecular Biology Unit, Department of Biochemistry, University College and Middlesex School of Medicine, Windeyer Building, Cleveland Street, London W1P 6DB, UK.

JAMES W. McCARRICK
The Wistar Institute, 3601 Spruce Street, Philadelphia PA 19104, USA.

AMY B. MacDERMOTT
Department of Physiology, Columbia University College of Physicians and Surgeons, 630 W168 Street, NY 10032, USA.

JOHN A. PIZZEY
Department of Anatomy and Human Biology, King's College, 26–29 Drury Lane, London WC2B 5RL, UK.

PIERRE ROUGET
Laboratoire de Biochemie Cellulaire, Collège de France, 11 Place Marcelin Berthelot, 75231 Paris cedex 05, France.

IRA SCHIEREN
Howard Hughes Medical Institute, Columbia University, College of Physicians and Surgeons, 701 W168 Street, NY 10032, USA.

BRUCE H. WAINER
Department of Pharmacological and Physiological Sciences, University of Chicago, 947 East 58th Street, Chicago, IL 60637, USA.

JOHN N. WOOD
Department of Neuroimmunology, Sandoz Institute for Medical Research, 5 Gower Place, London WC1E 6BN, UK.

JAMES R. WOODGETT
Ludwig Institute for Cancer Research, Courtauld Building, Riding House Street, London WC1, UK.

Abbreviations

ACh	acetylcholine
AChE	acetylcholinesterase
ADP	adenosine diphosphate
AMP	adenosine monophosphate
ATP	adenosine trisphosphate
BDNF	brain-derived neurotrophic factor
Bes	(N,N-bis[2-hydroxyethyl]-2-aminoethane sulfonic acid
BSA	bovine serum albumin
$[Ca^{2+}]_i$	free intracellular calcium concentration
CAM	cell adhesion molecules
CAT	chloramphenicol acetyl transferase
CD	Ca^{2+}-dependent
ChAT	choline acetyl transferase
CI	Ca^{2+}-independent
CIg	cold-insoluble globulin
CJM	cell junctional molecules
CMF	calcium- and magnesium-free Tyrode's solution
CNS	central nervous system
c.p.m.	counts per minute
CTP	cytidine triphosphate
CVR	current voltage relationship
DBH	dopamine-beta-hydroxylase
DEAE	diethyl-aminoethyl
DEPC	diethylpyrocarbonate
DIA	differentiation inhibiting activity
DMEM	Dulbecco's modified Eagle's medium
DMSO	dimethylsulphoxide
DNA	deoxynucleic acid
DTT	dithiothreitol
EDTA	ethylene diamine tetraacetic acid
EGF	epidermal growth factor
EGTA	ethylene glycol-bis(β-aminoethyl ether)N,N,N',N'-tetraacetic acid
EC	embryonal carcinoma
EIA	enzyme-linked immunoassay

Abbreviations

ELISA	enzyme-linked immunoassay
EMEM	Eagle's minimal essential medium
ES	embryonic stem cell
Fab	immunoglobulin variable region fragment
FCS	fetal calf serum
FGF	fibroblast growth factor
GABA	γ-amino butyric acid
GAP-43	growth-associated protein 43 kDa
GCT	germ cell tumours
GDP	guanosine diphosphate
GFAP	glial fibrillary acidic protein
GM-CSF	granulocyte/macrophage colony stimulating factor
GPI	glucose phosphate isomerase
GPPNHP	β-γ-imido guanosine triphosphate
GTP	guanosine trisphosphate
GTPγS	guanosine 5'-O'(3-thiotrisphosphate)
HAT	hypoxanthine aminopterin and thymidine
HBSS	Hepes buffered standard saline
Hepes	N-[2-hydroxyethyl] piperazine-N'-[2-ethanesulfonic acid]
HN	hippocampal hybrid cell lines
HMBA	hexamethylene bisacetamide
HPLC	high pressure liquid chromatography
HPRT	hypoxanthine phosphoribosyl transferase
5-HT	5-hydroxytryptamine
IMP	inosine monophosphate
InsP3	inositol tris phosphate
LIF	leukemia inhibitory factor
LTR	long terminal repeat
MARCKS	myristylated alanine rich C-kinase substrate
MBP	myelin basic protein
MCh	acetyl-beta-methylcholine
mCi	millicurie
MMTV	mouse mammary tumour virus
M.MuLV	Moloney murine leukemia virus
Mops	3-(N-morpholino) propane sulfonic acid
MPP$^+$	active metabolite of MPTP
MPTP	N-methyl-4-phenyl-1,2,3,6-tetrahydropyridine
mRNA	messenger ribonucleic acid
N-CAM	neuronal cell adhesion molecule
ND	rat dorsal root ganglion derived cell lines
NF	neurofilament
NFU	national formulatory units (trypsin measure)
NGF	nerve growth factor
NT-3	neurotrophin-3

Abbreviations

OCT-2	octamer-binding protein 2
ONPG	ortho nitrophenyl β-D-galactopyranoside
PAGE	polyacrylamide gel electrophoresis
PBS	phosphate-buffered saline
PCR	polymerase chain reaction
PEG	polyethylene glycol
PDBU	phorbol dibutyrate
p.f.u.	plaque-forming units
PHAP	phytohemaglutinin-P
PLL	poly-L-lysine
PMA	phorbol 12-myristic 13-acetate
PMT	photomultiplier tube
RA	all-trans-retinoic acid
RIA	radioimmunoassay
RNA	ribonucleic acid
RT	room temperature
SAM	substrate adhesion molecules
SDS	sodium dodecyl sulfate
SFM	serum-free medium
SSC	saline sodium citrate buffer
SN	septal hybrid lines
STE	saline tris EDTA buffer
ST	sterile tyrode's solution
SV40	simian virus 40
TBE	tris borate EDTA buffer
Tes	*N*-tris(hydroxymethyl)-2-aminoethane sulfonic acid
TLC	thin-layer chromatography
tRNA	transfer ribonucleic acid
UTP	uridine triphosphate

1

Neuronal hybrid cell lines: generation, characterization, and utility

BRUCE H. WAINER and ALFRED HELLER

1. Introduction

The nervous system comprises an extensive array of neuronal pathways that control and modulate virtually all bodily functions. This system develops from a simple neuroectodermal tube, and through a series of processes regulating cell division, commitment, migration, and differentiation, forms functional synaptic connections in a highly specific fashion. These processes involve internal cell programmes controlling stage-specific gene expression, cell–cell interactions, and chemical signalling. The result is an organ system of remarkable complexity, responsible for the full range of physiological and intellectual behaviour of higher organisms. A major challenge to neurobiologists is understanding the specific cellular and molecular interactions that mediate the formation and/or maintenance of functional connections between particular subsets of neurons and their target cells (1). This goal is made exceedingly difficult by the extensive heterogeneity of cell types and connections even within specific brain regions. For these reasons, a variety of primary cell culture techniques have been developed to provide an approach to simplification of the system for experimental purposes. These techniques are still limited by the cell heterogeneity of specific regions of brain and the difficulty of obtaining sufficient numbers of cells of a specific type for detailed biochemical and molecular biological studies. One approach that has been employed to circumvent these problems has been the utilization of clonal cell lines that exhibit neuronal features (2–4). For example, PC12 cells, arising from a rat pheochromocytoma, have been used in a variety of studies to elucidate the response to and the mechanism of action of nerve growth factor (NGF) (5, 6). Although this approach has provided a considerable amount of information, it is limited for the following reasons. First, these cell lines are derived from spontaneously arising tumours and therefore carry with them the inherently malignant nature of their cells of origin. Although a recent report has described the isolation of a human neural cell line from non-neoplastic tissue (7), the frequency of such spontaneous 'immortalization' events from primary neuronal cells is extremely low. Therefore, the possibility of obtaining cell lines of a specific neurochemical type is unlikely. Second, most of the cell lines currently available represent subclones of single tumours, arising most frequently in the peripheral nervous system, i.e. pheochromocytomas and neuroblastomas, and are therefore of limited usefulness for studying specific processes within particular CNS pathways.

Relatively little work has been carried out with respect to the establishment of permanent cell lines from specific brain regions that elaborate or respond to trophic signals which are involved in the establishment and maintenance of the synaptic circuitry of those regions. Two general strategies are available for engineering such cell lines. The first is the use of retroviral-mediated introduction (transduction) of oncogenes to 'immortalize' primary brain cells (Chapter 2 and refs 8 and 9). While this approach is useful for studying the properties of progenitor cells, it is of more limited utility as an approach to the study of differentiated cells. Retroviral transduction is only effective with mitotic cells since the retroviral DNA can only be inserted into the host genome during replication. In addition, once a cell is 'immortalized', it tends to remain locked within a particular developmental window, and in fact, this phenomenon has been exploited by immunologists to study the stages of lymphocyte differentiation (10, 11). Therefore, while viral gene transduction might yield cell lines for the study of early stages of neuronal development, it is less likely to provide cell lines that express the phenotypic repertoire of mature neurons which are almost invariably post-mitotic.

A second approach has employed somatic cell fusion techniques in which primary brain cells are fused to a neuroblastoma cell line by exposure to polyethylene glycol (12). The fusion technique allows one to 'immortalize' cell populations that are post-mitotic and therefore more likely to express highly differentiated neuronal phenotypes. The purpose of the present chapter is to present the methodological considerations that are essential in applying the somatic cell fusion approach to the immortalization of CNS cells, and to discuss the properties of hybrid cell lines that have been generated from neurons of the septohippocampal and nigrostriatal pathways (*Figure 1*). While

---

Figure 1. Schematic illustration of the generation of somatic hybrid cell from the septohippocampal and nigrostriatal pathways. **1** Primary cells are removed from either embryos or post-natal animals at an age when the neurons of interest have become post-mitotic or have developed the appropriate neuronal phenotypes of interest. **2** and **3** Schematic illustrations of the septohippocampal (S, septal area; and HI, hippocampus) and nigrostriatal (SN, substantia nigra; and CS, corpus striatum) pathways. **4** Primary brain cell suspension following trypsin treatment and mechanical dissociation. **5** Buoyant density gradient employed for harvesting of post-natal primary brain cells bottom to top as follows: *Black* = 8% BSA–10% sucrose in DMEM containing large clumps of cells and

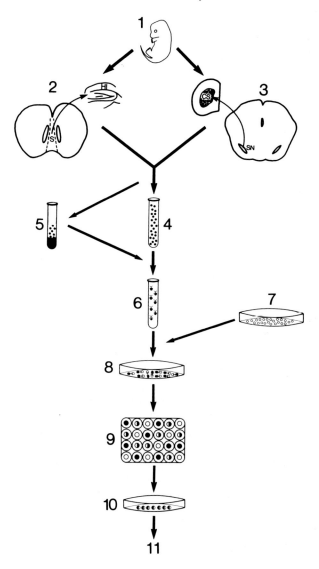

debris after centrifugation (yellow); *Dots* = middle zone containing 3.75% BSA in DMEM and enriched for viable primary brain cell following centrifugation (orange); *Grey* = cell suspension in DMEM which is layered on before centrifugation and which contains subcellular debris following centrifugation (red). **6** Suspension of primary brain cells that have been treated with PHA-P. **7** Monolayer culture of N18TG2 neuroblastoma cells. **8** Co-culture of N18TG2 and primary brain cells in culture plate that is treated with PEG. **9** Microcultures in HAT-containing media consisting of unfused neuroblastoma (*white*) and primary brain cells (*black*), which exhibit no further growth, and hybrid cell lines (*white/black*) which appear as visible colonies in 2–3 weeks. **10** Expanded hybrid cell line colonies which are subcloned; **11** screened for appropriate neurochemical markers.

no current 'cell immortalization' strategy is devoid of limitations, the somatic cell fusion technique makes available large numbers of brain-region-specific clonal cells for cellular and molecular studies of specific neural circuits.

2. Methodology

The technique of somatic cell fusion has been widely used to study a variety of cellular and genetic questions (13). Perhaps one of most noteworthy applications has been the generation of lymphoid hybridoma cell lines to produce monoclonal antibodies (14). In the nervous system, somatic cell fusion has been applied to the development of several cell lines derived from sympathetic neurons (15) or dorsal root ganglion cells (16). We have exploited the cell fusion approach, illustrated schematically in *Figure 1*, because of its potential for immortalizing central neurons that are post-mitotic and therefore committed to a particular neurochemical phenotype as well as neuroanatomical pathway (12, 17–19). This section considers technical aspects related to preparation of primary cells, the fusion partner cell line, and the fusion protocol itself.

2.1 Dissection and dissociation of primary mouse brain cells

Primary brain cells represent a source of neuronal cells that can be targeted for immortalization. It is essential that the brain region of interest be readily identifiable under a dissecting microscope, particularly if embryonic animals are employed. It is also important that the embryonic stage at which particular neurons are born, i.e. become post-mitotic and already committed to a particular neurochemical phenotype, is known. Based on the [^{3}H]-thymidine technique, a considerable 'birth-dating' literature is presently available for many brain regions. For example, neurons of the septal regions undergo their terminal mitoses within a developmental window of embryonic days 14–16 (E14–16) (20–22).

The following procedure, modified from Hemmendinger *et al.* (17, 23) for primary cultures, has been employed for preparing suspensions of primary embryonic brain cells for fusion purposes. It is based on a typical experiment in which 50 embryonic brains are used.

Protocol 1. Dissection and dissociation of embryonic brain cells

1. On the appropriate day of pregnancy, day 0 being the day on which a vaginal plug is first detected, kill the female mice by cervical dislocation, remove the uteri and place in a Petri dish filled with sterile Tyrode's solution (ST) containing:

	g/litre
• sodium chloride	8.0
• potassium chloride	0.2

- calcium chloride 0.2
- magnesium chloride 0.1
- sodium phosphate monobasic 0.05
- sodium bicarbonate 1.0
- glucose 1.0

2. Remove the embryos from their uteri, stage them according to Gruneberg (24) and discard embryos which are at an inappropriate stage.
3. Dissect the regions of interest using fine iris knives under a dissecting microscope.
4. Transfer the tissues to a 10-ml conical centrifuge tube and rinse 3–4 times with 2 ml of Ca^{2+}-Mg^{2+}-free Tyrode's solution (CMF). Place the tubes under an atmosphere of humidified 5% CO_2–21% O_2–74% N_2 to adjust the pH to 7.4, and incubate at 37°C with 2 ml of CMF for 20 min.
5. Decant the CMF and replace with a solution of CMF containing 0.1–0.67% trypsin depending on the brain area involved (23, 25) and allow the tissue to incubate for an additional 35 min.
6. Remove the trypsin solution and wash the tissue carefully, 3–4 times with CMF and once with the initial culture medium consisting of:
   - Eagle's basal medium 89 ml
   - fetal bovine serum 10 ml
   - penicillin–streptomycin solution[a] 1 ml
   - deoxyribonuclease I 25 mg
7. Dissociate the tissue in 1 ml of initial culture medium by repeated gentle flushing through a fine-bore Pasteur pipette.
8. Examine a sample of the resulting cell suspension microscopically to ensure complete dissociation.
9. Determine the number of cells by counting with a haemacytometer.[b]

[a] The penicillin–streptomycin stock solution used contains 5000 units penicillin/ml and 5000 µg streptomycin/ml.
[b] Typically, 5.0×10^6 cells are obtained from the rostral mesencephalon of ten E14 embryos or the corpus striatum of five E14 embryos. Similar numbers of cells are obtained from the septal region of ten E15 embryos or the hippocampus of seven E15 embryos. Viability of the cell suspensions should range from 75 to 100%, as determined by trypan blue exclusion.

This procedure works well with tissues derived from embryonic brains, but is inadequate for postnatal tissue. However, as described below, several fusions have been successfully performed on cells obtained from post-natal animals (18, 19). Protocol 1 above does not provide sufficient numbers of viable cells for these studies, since a considerable amount of debris is present in the suspensions. In order to eliminate debris and increase the yield of viable cells, the procedure in *Protocol 2* is utilized. The conditions described here are based on the harvesting of post-natal day 21 (P21) septal cells from 29 brains (18).

Protocol 2. Dissection and dissociation of post-natal brain cells

1. The dissection of post-natal brain tissue is carried out in Dulbecco's modified Eagle's medium (DMEM). The dissection is performed on ice and all solutions are cooled to 4°C, except trypsin.
2. Wash the tissue with CMF and incubate in trypsin (see *Protocol 1*).
3. Dissociate the tissue by passing gently (12 times) through a siliconized, autoclaved Pasteur pipette with a fire-polished tip. Allow the debris to settle by gravity for 1 min, and remove the cell suspension (supernatant).
4. Resuspend the undissociated tissue (precipitant) in additional medium and carry through two additional cycles of mechanical dissociation. Pool the three supernatants (final volume of approx. 15 ml) and discard the remaining undissociated tissue.
5. Centrifuge the pooled cell suspension at 33 g for 4 min through a density gradient composed from bottom to top as follows:
   (a) 8% bovine serum albumin (BSA) in DMEM containing 10% sucrose (w/v)
   (b) 3.75% BSA in DMEM
   (c) the cell suspension
6. After centrifugation, each zone is clearly demarcated by the presence of phenol red indicator in the DMEM. The upper zone (DMEM; red) contains subcellular debris while the lower zone (8% BSA–10% sucrose; yellow) contains tissue membranes and cell clumps. The central zone (3.75% BSA; orange) and its lower interface is enriched for single cells which are removed by aspiration.
7. Collect cells in this enriched cell suspension by centrifugation at 1138 g for 6 min and use for somatic cell fusion with the N18TG2 neuroblastoma.[a]

[a] Use of this gradient and the modified dissociation protocol yields approx. 1.7×10^6 cells per septum with a 78% viability, while the conventional dissection protocol yields 100 times fewer cells per septum at about half the viability (18).

2.2 Neuroblastoma fusion partner

In all of the hybrid cell lines generated, the cell line used as a fusion partner has been the murine neuroblastoma line N18TG2 which is deficient in hypoxanthine phosphoribosyltransferase (HPRT:IMP:pyrophosphate phosphoribosyltransferase, EC 2.4.2.8) and can be selected against in culture medium containing hypoxanthine, aminopterin, and thymidine (HAT) (15, 26, 27). N18TG2 cells were chosen because they are embryologically related to neuronal cells, and thus likely to permit the expression of neuron-specific traits (28, 29). The murine origin of both the neuroblastoma line and the normal parent cells minimizes chromosomal loss which often occurs in xenogeneic fusions (30, 31).

Protocol 3. Preparation of neuroblastoma (N18TG2) cells for fusion

1. In order to avoid HPRT reversion in the N18TG2 cells, culture the cells in tissue culture plastic plates in 0.1 mM 6-thioguanine in Matalon's modified Eagle's medium (EMS) (Gibco) with 10% (v/v) fetal calf serum (FCS) at 37°C in 10% CO_2 which selects against such revertants.
2. Seven days before a fusion, subculture the N18TG2 cells in 6-thioguanine-free medium.
3. Four days later, replate the cells in 100 mm tissue culture plastic plates at a density of 15 000 cells per cm^2.
4. On the day of fusion, incubate the neuroblastoma cells for 10 min, at 37°C, in 0.01% trypsin in CMF. After the incubation period, add an equal volume of FCS-containing medium and harvest the cells with a Pasteur pipette by mechanical agitation.
5. Centrifuge the cell suspension at 200 g for 10 min, and resuspend in medium. Plate the cell suspension on 60-mm Falcon tissue culture plates at a density of 2.5×10^6 cells per plate.

2.3 Cell fusion

The fusion procedure described by Fournier (32) was modified to enhance fusion of primary brain cells with N18TG2 cells (12, 17, 18). This procedure works equally well with embryonic or post-natal cells. We found that the use of phytohemagglutinin-P (PHA-P) and the duration of exposure to polyethylene glycol (PEG) effect the efficiency of fusion. Primary cells do not adhere well to N18TG2 cells when standard monolayer fusion techniques are used. The use of PHA-P promotes the adherence of essentially all primary cells to the N18TG2 monolayer, and dramatically improves the efficiency of fusion. Control plates of primary cells or N18TG2 cells alone, treated in a similar manner, are also prepared in parallel with the experimental plates.

Protocol 4. Fusion of primary cells with N18TG2 cells

1. Resuspend the primary brain cells at 10^7 cells/ml (see *Protocols 1* and *2*) in a solution of 1 µg PHA-P/ml DMEM.
2. Wash the plated N18TG2 cells gently with serum-free medium, add 2.0 ml of the PHA-P-treated primary cell suspension (step 1) to each plate and in-

Protocol 4. *Continued*

cubate the cells for 15 min at 37°C. The final ratio of primary brain cells to tumour cells is 10:1. After incubation, the plate is examined under a microscope to assure that all of the primary cells adhere to the N18TG2 monolayer.

3. Aspirate the PHA-P solution and apply 1.5 ml PEG solution [50/50 (w/w) solution of PEG, mol. wt. 1000 (Koch-Light 1000, Research International), in serum-free medium, at pH 7.6].
4. After 60 sec,[a] aspirate the PEG solution and rinse the plate quickly three times with serum-free medium. Add a volume of 3.5 ml of medium with 10% FCS and incubate the plates overnight at 37°C in 10% CO_2.
5. After 24 h, change the medium to HAT medium in order to select against unfused HPRT-deficient N18TG2 cells (26, 27). The HAT medium contains:
   - DMEM   90 ml
   - FCS   10 ml
   - hypoxanthine   100 µM
   - aminopterin   0.4 µM
   - thymidine   16 µM
6. Harvest the fusion products with a Pasteur pipette by mechanical agitation and plate in HAT medium at 50 000–100 000 viable cells, as established by trypan blue exclusion, per 35-mm tissue culture well (Falcon). Alternatively, cells can be plated in 6.4-mm microculture plates at a density of 100 000/well. Change the medium every 3 days.
7. Examine the culture plates (within 2–3 weeks) under the microscope for the appearance of colonies. Individual colonies can be isolated and expanded as follows:
   (a) For the 35-mm plates, individual colonies are isolated using cloning cylinders, and each colony is replated in a 16-mm well, in medium containing HAT and 10% FCS.
   (b) For the microculture plates, single colonies can be transferred and expanded in a similar fashion as for 35-mm plates.
   (c) Where more than one colony is present, the hybrid cells can be cloned by limiting dilution (12).
8. Expand the colonies and screen for relevant characteristics such as morphology, cytoskeletal, and neurochemical markers.
9. Culture cell lines of interest in DMEM with 10% FCS.

[a] The duration of exposure to PEG is critical (33). Exposure to PEG for longer than 50–60 sec under these conditions results in marked cell death, probably due to the cytotoxicity of PEG itself. The efficiency of fusion declines with shorter exposure times.

2.4 General characterization of hybrid cell lines

Before evaluating the various cell colonies for specific neurochemical properties, it is important to verify the hybrid nature of the cell lines generated. The fact that the cells are capable of growth in HAT medium strongly suggests that the cells have retained the growth characteristics of the parent neuroblastoma as well as the normal HPRT gene from the parent primary cells. However, it is possible that HPRT revertants have developed although the expected frequency would be very low (less than 10^{-7}) (15). We have never observed any growth in control fusion plates containing N18TG2 cells alone. In control primary cell plates, the only growth observed consists of slowly dividing colonies with glial morphologies. The hybrid nature of the cells obtained following fusion is confirmed both by karyotype analysis and by determination of enzyme polymorphisms.

Karyotype analysis is performed as previously described (12, 34–36). For each cell line, ten Giemsa-stained metaphase chromosome spreads are photographed and the total number of chromosomes counted. Cells from C57BL/6 mice, which constitute the source of primary parent cells, have 40 chromosomes per cell, all acrocentric or telocentric (36). N18TG2 cells have a mean of 79 ± 2 ($n = 10$) chromosomes per cell with large metacentric chromosomes present (mean = nine per cell) (12). All HAT-resistant cell lines which have been examined to date contained more chromosomes than either parental cell type and have metacentric chromosomes (12, 17–19). Since N18TG2 cells contain metacentric chromosomes, but C57BL/6 cells do not, at least one parent of each hybrid line is an N18TG2 cell (12, 15).

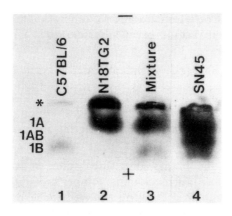

Figure 2. Electrophoretic analysis of GPI isozymes in SN45, a post-natal septal hybrid cell line. C57BL/6 and N18TG2 cells express only the IB or IA homodimer of GPI (*lanes 1 and 2*). A mixture of C57BL/6 and N18TG2 cell extracts expresses both homodimers but no heterodimer (*lane 3*). An extract from the hybrid cell line (*lane 4*) expresses a heterodimeric form of GPI (IAB) as well as both homodimers. The asterisk indicates the origin of the gel. From Lee *et al.* (18), by permission.

Neuronal hybrid cell lines: generation, characterization, and utility

The presence of functional chromosomes from both parents is also documented by analysing glucosephosphate isomerase (GPI; EC 5.3.1.9) isozyme expression (12, 18, 37, 38). This enzyme normally exists as a homodimer. It had been shown previously that A strain mice (from which N18TG2 cells are derived) and C57BL/6 mice (the source of the primary cells) express electrophoretically distinct variants of GPI (15). Furthermore, cells from animals heterozygous for GPI express the variant of each parent as well as an intermediate form (39, 40). Neuronal hybrid cells express the GPI electrophoretic variants of both parental strains, as well as an intermediate form, which does not appreciably migrate under these conditions (*Figure 3*). These findings demonstrate that these cells are hybrids derived from the fusion of C57BL/6 primary brain cells and N18TG2 neuroblastoma cells. The GPI analysis indicates that the hybrid cells are indeed capable of expressing genes from each of the parent cell types.

Protocol 5. Analysis of glucosephosphate isomerase isozyme expression

1. Preparation of extracts from hybrid cell lines and N18TG2 cells
   (a) Grow the cells in DMEM with 10% FCS to approximately 95% confluency on tissue culture plastic plates.
   (b) Rinse the culture plates with calcium- and magnesium-free Hanks' balanced salt solution (CMF-HBSS; Gibco) at 37°C, harvest the cells by mechanical agitation, and centrifuge at 1000 g for 5 min at 4°C.
   (c) Resuspend the cells in ice-cold HBSS and recentrifuge.
   (d) Resuspend the pellet in an equal volume of extraction buffer (5 mM Tris, pH 7.5, 1 mM disodium EDTA, and 2.0% Triton X-100; Innovative Chemistry) for 30 min on ice.
   (e) Centrifuge at 2000 g for 10 min at 4°C.
   (f) Store the supernatant in liquid nitrogen.
2. Preparation of C57BL/6 adult mouse brain extract:
   (a) Remove the brain, cut in half, quickly freeze in liquid nitrogen, and store at −80°C.
   (b) Thaw half of the brain, rinse three times in ice-cold HBSS, and homogenize in a TenBroeck tissue grinder in 360 µl of ice-cold extraction buffer containing 25 µg/ml deoxyribonuclease I (3381 units per milligram dry weight; Worthington Biochemical Corporation).
   (c) Centrifuge the homogenate at 2000 g for 10 min at 4°C.
   (d) Store the supernatant in liquid nitrogen.
3. Electrophoresis is performed at 4°C for 50 min at a constant voltage of 200 V. Electrophoresis gels contain 1% agarose and 5% sucrose in an

aminomethylpropanol buffer at pH 8.9. The electrode buffer is sodium alkaline barbital buffer (50 mM, pH 8.6) (Innovative Chemistry). Bands for GPI activity are visualized by staining the gel at 37°C using commercially available GPI substrate (Innovative Chemistry).

3. Septal and hippocampal hybrid cell lines

Neurons of the septal region, some of which are cholinergic, and their axonal projections to target cells in the hippocampus comprise the septohippocampal pathway which plays an essential role in cognitive processes (41). The anatomy and function of this pathway have been studied extensively (42, 43), and its selective vulnerability in Alzheimer's disease is well-recognized (44). In particular, it has been employed as a model system to study mechanisms of neural injury (41), plasticity (45) and the function of neural growth factors (46). Although the pathway is well-defined anatomically, the cell types involved are complex and a growing number of trophic factors have been reported to exert various effects during development, ageing, and following injury (47–51). In order to better define the properties of various cell types that comprise the septohippocampal pathway and the actions of various trophic factors, we utilized the somatic cell fusion technique to immortalize neurons of the septal region and their target cells of the hippocampus (12, 17–19).

3.1 Septal cell lines

Septal hybrid cell lines were initially generated from embryonic brain cells and subsequent modification of the cell harvesting procedure enabled the establishment of cell lines from post-natal/young adult brains. To date, 60 septal hybrid lines (SN) have been generated from E14, 15, 18, and P21 tissues (12, 17, 18). The neuronal phenotypes of these cell lines have remained stable under continuous culture conditions for over 1 year. The cells are grown in tissue culture plastic plates, in DMEM supplemented with 10% FCS. Since chromosomal instability can occur, we routinely subclone cell lines of interest and screen for the particular phenotypic characteristics for which we originally selected. This approach has ensured the continued stability of phenotypic expression in these lines. Samples of all cell lines and subclones are also stored by freezing in liquid nitrogen in medium containing dimethylsulfoxide.

Cell lines that have been studied in greatest detail were initially selected on the basis of morphology (*Figure 3*). For example, colonies exhibiting reasonable generation times of 24–72 h with 'neurite-like' processes were isolated for expansion, subcloning, and further analysis, while slowly dividing colonies with flat 'glial-like' appearances were not examined further. Typically, the

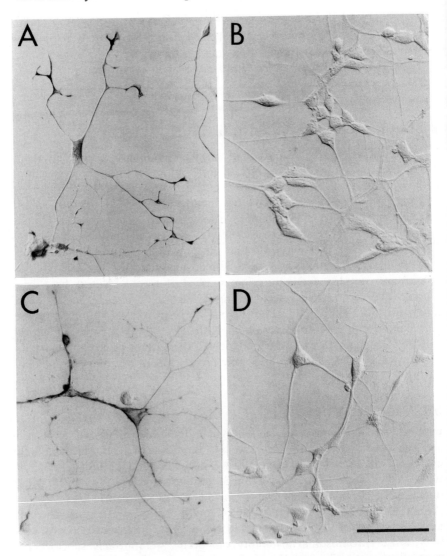

Figure 3. Neuronal morphology and neurochemical features of the postnatal septal cell line, SN56. **A**: Immunocytochemical localization of neurofilament proteins employing a mixture of antibodies against low, middle, and high molecular weight forms (52). **B**: Absence of immunoreactivity following incubation with antibodies against the astroglial intermediary filament, GFAP. **C**: Expression of acetylcholinesterase reaction product in the presence of the butyrylcholinesterase inhibitor, iso-OMPA (12). **D**: Absence of acetylcholinesterase reaction product in the presence of the specific acetylcholinesterase inhibitor, BW284C51 (12). Calibration bar: 100 μM.

cell lines of interest have somata and processes which are highly refractile. Ultrastructurally, the processes display neuronal intermediate filaments and other features of developing neurons such as abundant growth cones (17). Immunocytochemical and immunoblotting studies have confirmed the presence of neurofilaments (*Figure 3A*), and demonstrate that the cytoskeletal features of some lines are sufficiently differentiated to express the low-molecular-weight form as well as differentially phosphorylated isoforms of the middle and high-molecular-weight subunits. This cytoskeletal array is characteristic of mature neurons (52, 53). Septal cell lines cultured continuously for over 1 year still express all neurofilament subunit isoforms and presumably the various kinases and phosphatases required for regulation of these neurofilament phosphorylation states. Intermediate filaments characteristic of non-neuronal cells, such as cytokeratins or GFAP, are not detected in SN lines expressing neurofilaments (*Figure 3B*) (12, 18, 52). When compared to the embryonic cell lines, the post-natal lines are more highly differentiated, particularly with respect to expression of the phosphorylated neurofilament isoforms.

One important question relating to these cell lines is their capacity to develop neurotransmitter-associated phenotypes typical of septal neurons. We screened initially for cholinergic features since acetylcholine (ACh) is a major transmitter expressed in the septohippocampal projection and cholinergic neurons are abundant in the medial septal/diagonal band region. Measurements of choline acetyltransferase (ChAT) activity, the biosynthetic enzyme for ACh, revealed a significant number of positive cell lines (*Figure 4*) (12, 17). In contrast, the parent N18TG2 neuroblastoma has barely detectable ChAT levels. As with cytoskeletal features, the post-natal cell lines express higher ChAT levels on average than the embryonic cell lines (*Figure 4*) (17). In addition, many of the cell lines also express the degradative enzyme for ACh, acetylcholinesterase (AChE) (*Figure 3*) (12). The specificity of this enzyme activity is demonstrated by its insensitivity to the butyrylcholinesterase inhibitor, iso-OMPA (*Figure 3C*), and its sensitivity to the specific AChE inhibitor, BW284C51 (*Figure 3D*). The cholinergic properties of the post-natal cell line SN56.B5.G4 have been further examined in collaboration with Dr J. K. Blusztajn at Boston University, who has found that this cell line expresses three additional 'cholinergic' properties (54, 55):

- the capacity to synthesize ACh;
- a high-affinity sodium-dependent choline uptake system; and
- the capacity to release ACh following depolarization with high potassium.

These findings are of considerable interest since there have been no previous reports of cell lines which express all components of normal cholinergic neurochemical function.

Another important component of the characterization of any 'neuronal'

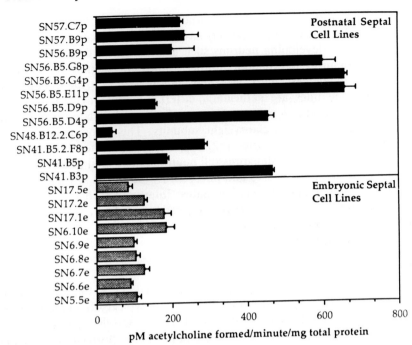

Figure 4. Expression of choline acetyltransferase activity in septal hybrid cell lines. On average the embryonic cell lines (*grey*) contain lower levels of enzyme than the post-natal lines (*black*). From Wainer et al. (60, 77), by permission.

cell line is an examination of the expression of excitable properties. A preliminary electrophysiological study of the properties of an embryonic cell line, SN6.10.2, has been performed in collaboration with Drs D. A. Brown and J. A. Sim, University of London (unpublished observations). Following differentiation with retinoic acid, these cells exhibit resting membrane potentials of −40 to −60 mV. The cells also display sodium and potassium currents sufficient for generation of action potentials. Similar properties have been observed in the post-natal line SN56 by Dr N. L. Harrison, University of Chicago (unpublished observations).

Primary septal cholinergic neurons express the NGF receptor and are responsive to NGF in terms of elevation of ChAT activity and prevention of cell death following axotomy (46, 47, 56). In preliminary studies, selected hybrid septal lines were screened for either morphological or neurochemical (ChAT activity) responses to NGF (17, 18). While these particular cell lines did not respond to NGF, it should be noted that the assays were performed under culture conditions that promote cell proliferation (i.e. 10% serum supplementation). More recently, we have observed that some of the septal cell lines expresses low affinity NGF receptor mRNA as well as high and low affinity NGF-binding sites (Dr H. C. Palfrey, University of Chicago, unpublished observations), but nevertheless do not respond to NGF. It may be that the cell lines lack an important signal transduction mechanism for eliciting an NGF effect.

The recent identification of the 'trk' family of protooncogenes which function as high affinity neurotrophin receptors has provided further insights into the active receptor complex (57). It will therefore be important to evaluate the septal cell lines for *trk* expression. A recent report by Tabira and co-workers has described trophic responses of septal cholinergic neurons to interleukin 3 (IL3) (58). These investigators found that one of our septal cell lines, SN6.10.2.2, also responds to IL3 with significant elevations of ChAT activity. More recently, both primary septal cells and this septal cell line, SN6.10.2.2, were found to respond similarly to another cytokine, granulocyte–macrophage colony stimulating factor (GM-CSF) (59). These findings suggest that the existing septal cell lines will be quite useful for studying neuronal responses to specific varieties of trophic signals.

3.2 Hippocampal cell lines

To date, 48 hippocampal hybrid cell lines (HN) have been generated from E18 and P21 brain cells (19). Some of these lines exhibit neuronal morphologies and cytoskeletal features, as well as excitable properties typical of hippocampal neurons. The HN cell lines have not been extensively studied with respect to neurotransmitter-associated properties. These cell lines have been examined from the perspective of the elaboration of trophic signals that may influence the development of septal neurons or differentiation of SN cell lines. Since NGF is a putative trophic factor for the septohippocampal pathway, HN cell lines were screened for expression of NGF mRNA and NGF protein. As illustrated in *Figure 5*, several of these cell lines do express NGF. On average, the post-natal HN lines tend to express higher levels of NGF protein than the embryonic lines.

The HN cell lines were also examined for expression of non-NGF trophic factors. A microculture bioassay was developed to screen either conditioned media or subcellular fractions of hippocampal cell lines for the presence of molecules which might elevate ChAT activity in primary septal cholinergic neurons (19, 60). To date, 22 HN cell lines, as well as N18TG2 cells have been screened. Most of the lines have little effect (less than 76% increase in ChAT activity). However, either conditioned medium or a crude membrane extract from one line, HN10, promotes a fourfold increase in ChAT activity.

This activity is not neutralized by antibodies against NGF, is not attributable to either epidermal growth factor or basic fibroblast growth factor, but is inactivated by proteolysis. The recent identification of BDNF and NT-3 as members of a family of NGF-related trophic factors, the neurotrophins (61, 62), raises the possibility that one or a combination of these factors may be responsible for the HN10 activity. The HN lines are presently being screened for BDNF/NT-3 expression.

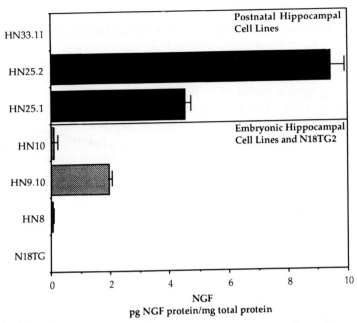

Figure 5. Expression of NGF protein in hippocampal hybrid cell lines. On average the embryonic cell lines (*grey*) contain lower levels of NGF than the post-natal lines (*black*). From Lee et al. (19), by permission.

4. Mesencephalic and striatal cell lines

The dopaminergic neurons of the substantia nigra and their telencephalic projections (i.e. basal ganglia/neocortex) have been the subject of intense investigation for more than three decades. These neurons appear to subserve a variety of physiological functions ranging from motor function to fundamental mental processes whose dysfunction can result in diseases as disruptive as schizophrenia (63). Particular attention has been focused on the development, anatomy, physiology, and functional importance of the dopaminergic nigrostriatal projection whose degeneration appears to play a pivotal role in the pathogenesis of Parkinson's disease (64). An understanding of the trophic interactions which modulate the establishment and maintenance of these essential central projections is therefore a subject of considerable interest particularly in light of the recent findings of the effects of BDNF on dopaminergic neuronal survival in culture (50, 65).

As described here, the availability of immortalized cell lines provides an approach to the examination of many of the issues involved in trophic regulation of development, neuronal specificity, and survival. The potential of this approach is well-illustrated by the properties possessed by the embryonic mesencephalic dopaminergic and striatal neurons which we have obtained to date by the somatic cell fusion technique. These cell lines would appear to provide a model for studying the role of chemotrophic signals in the establishment of specific dopaminergic connections, as well as a variety of other neuronal functions including catecholamine storage and synthesis, receptor characterization, and neurotoxicity.

4.1 Mesencephalic cell lines

Using the cell fusion protocol described above with E14 primary mesencephalic cells (*Figure 1*), a hybrid cell line (MN9D) has been generated which exhibits many of the characteristics of primary dopaminergic neurons (34). This cell line, in contrast to its neuroblastoma partner, contains substantial levels of catecholamines, including dopamine and norepinephrine, sufficient to permit the visualization of the cell bodies and processes of these cells by histofluorescence techniques as seen in *Figure 6A* and by immunocytochemical visualization of tyrosine hydroxylase as seen in *Figure 6B*. The cell line produces and releases sufficient dopa and dopamine so that these compounds are detectable in the culture media in which the cells are grown. The cells are quite stable and have been found to maintain catecholamine synthesis and release for up to 8 months in continuous passage and similar morphology and neurochemistry after more than 2 years of storage in liquid nitrogen. In

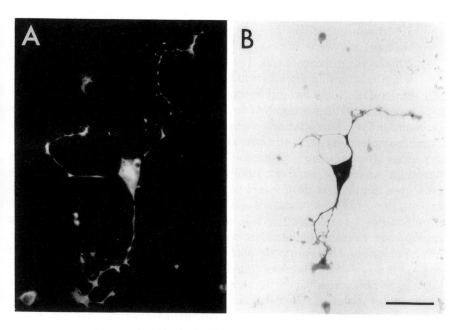

Figure 6. Neuronal morphology of the embryonic mesencephalic cell line MN9D in monolayer culture and processed to reveal (**A**) endogenous catecholamine histofluorescence and (**B**) tyrosine hydroxylase immunoreactivity. Calibration bar: 50 μM.

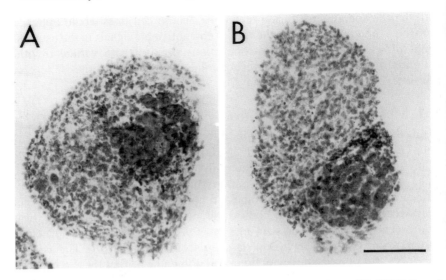

Figure 7. Representative cresyl violet stained tissue sections from (**A**) MN9D-Target aggregates and (**B**) MN9D-Non-target aggregates. MN9D hybrid cells can be distinguished from primary cells by their large size and are present in both types of aggregates. Calibration bar: 100 μM.

contrast to neuroblastoma cells, the MN9D cell line shares with primary embryonic neurons an ability to aggregate in rotational tissue culture (*Figure 7*) (34). This latter property makes it possible to examine a number of issues of cell–cell interaction between hybrid cell lines and primary cells (see below).

The MN9D cell line has interesting neurochemical features that may provide insights as to the ontogeny of dopamine synthesis in primary dopaminergic neurons (34). This cell line, in contrast to primary dopaminergic neurons, cannot completely convert all of the available dopa to dopamine. Specifically, the levels of dopa, the precursor of dopamine, are at about 50% of the dopamine levels. In addition, norepinephrine, a product of dopamine is present at about 30% of the level of dopamine. This neurochemical profile reflects the embryonic status of the hybrid cells since the developmental expression of tyrosine hydroxylase would appear to precede that of dopa decarboxylase throughout fetal and neonatal development (66). The presence of norepinephrine, not normally present in mature dopaminergic neurons, is due to the fact that these cells contain the enzyme dopamine-beta-hydroxylase (DBH) (Heller, unpublished results). The presence of DBH may reflect a transient expression of this enzyme in fetal cells since it is thought to be coded by the same or linked genes to that of tyrosine hydroxylase (67).

The MN9D cell line has a number of other properties which suggest it can serve as a useful model for examining mesencephalic dopaminergic neuronal function. Under whole-cell voltage-clamp conditions these hybrid cells are capable of generating depolarization-induced action potentials and voltage-sensitive sodium currents which are indistinguishable from those seen in rodent brain (34). Of particular interest is the finding that the MN9D line is considerably more sensitive than PC12 cells, derived from adrenal medullary pheochromocytoma, to the depleting actions of MPP$^+$, the active metabolite of N-methyl-4-phenyl-1,2,3,6-tetrahydropyridine, MPTP (*Table 1*). MPTP is a substance that has been shown to produce neurotoxicity to dopaminergic neurons in a variety of species including humans, mice, and monkeys (34, 68). Preliminary studies suggest that low concentrations of MPP$^+$ elicit cell death in the MN9D cell (69). These findings suggest that the MN9D line may provide a useful monoclonal hybrid cell model for studying mechanisms of neurotoxicity.

Table 1. Effect of MPP$^+$ on dopamine levels[a] in MN9D and PC12 cells

MPP$^+$ concentration	MN9D	PC12
0 M	101.4 ± 1.8	353.2 ± 25.5
10^{-6} M	43.1 ± 1.5	413.9 ± 11.8
10^{-5} M	33.6 ± 3.4	348.1 ± 16.5
10^{-4} M	15.1 ± 0.7	258.6 ± 11.8
10^{-3} M	9.4 ± 0.4	60.3 ± 2.5

[a] Dopamine levels are expressed in ng/mg protein as the mean ± the standard error of the mean ($n = 5$). This data was presented previously in Choi *et al.* (34), but was expressed as percentage of control.

The somatic cell fusion technique depends for its utility on an admixture of genetic material from both partners in the fusion, the neuroblastoma line and primary brain cells. In this situation it is quite likely that the majority of cell fusion products formed contain varying amounts of genetic material from each partner and thus may only partially express the phenotypic characteristics of a particular primary brain cell. An interesting illustration of this phenomenon is the cell line MN9H, which was subcloned from the same fusion which produced the MN9D line. This line, like MN9D, is capable of synthesizing dopa, dopamine, and norepinephrine, but contains approximately equivalent amounts of the catecholamines, dopamine, and norepinephrine, in contrast to MN9D where dopamine predominates (34). In the media in which the MN9H cells are grown, however, approximately 7–8 times the amount of dopamine is present as compared to media from MN9D cells (34). This result suggests that the MN9H cells exhibit a deficit in dopamine storage. The two lines, MN9D and MN9H, therefore represent a model system for examination of catecholaminergic storage function in well-defined clonal cell lines.

If neuronal hybrid cell lines are to be of use with regard to issues of cell–cell interaction and particularly, trophic interactions between brain cells, the lines should exhibit the capacity to distinguish between primary target and non-target cells. In the case of the MN9D cell, which is derived from mesencephalic dopaminergic neurons, one would expect a differential response to cells of corpus striatum or neocortex (areas to which the dopaminergic neurons project) as compared to optic tectum or thalamus (anatomic subdivisions of brain to which primary dopaminergic neurons do not project). Due to their ability to aggregate, as described above, it is possible to coaggregate these immortalized cells selectively with primary cells from various subdivisions of brain and thus permit the examination of selective cell–cell interactions. For this purpose, 2000 MN9D cells were coaggregated with 8 million cells of either corpus striatum, frontal or occipital cortex (target areas for dopaminergic neurons) or with 8 million cells of the optic tectum or thalamus (non-target areas). Following seven days in culture, in the presence of n-butyrate to suppress overgrowth of the hybrid cells, the aggregates were harvested. Analysis of cresyl violet stained sections revealed that viable hybrid cells were present, irrespective of the presence of target (*Figure 7A*) or non-target cells (*Figure 7B*). However, examination of catecholamine histofluorescence revealed a striking difference between the types of coaggregates: MN9D-target coaggregates exhibited highly fluorescent MN9D cells characteristic of the presence of catecholamines; while in MN9D-non-target coaggregates, no catecholamine histofluorescence was observed (70). Further analyses revealed that the loss of fluorescence in the non-target coaggregates was due to a marked difference (approximately 12- to 20-fold depending on the target area involved) in dopamine content between the coaggregates made with target cells as compared to non-target cells. This striking result appears on the basis of preliminary studies (H. Choi, unpublished observations) to be secondary to a down-regulation of mRNA for tyrosine hydroxylase, the rate limiting enzyme in catecholamine biosynthesis.

These findings on the selective effect of dopaminergic non-target cells on the mesencephalon-derived dopaminergic cell line MN9D, suggests that the immortalized neuronal hybrid cell lines derived from specific brain regions by somatic cell hybridization may serve as model systems for the examination of positive and negative signals involved in the elaboration of the complex and highly selective patterning of axonal connections in the intact brain.

4.2 Striatal cell lines

Cloning and characterization of hybrid cell lines from the corpus striatum, a principal target cell population for dopaminergic neurons, has not yet progressed to the extent described for the dopamine-containing MN9D line.

Studies on cells obtained from the somatic cell hybridization of E18 corpus striatum cells with N18TG2 neuroblastoma cells, however, suggest that the hybrid immortalization technique will prove useful in this case as well. Preliminary studies carried out in collaboration with Dr Bruce Perry at the University of Chicago have utilized these cell lines to study dopamine receptors. The hybrid corpus striatum cells obtained express D_1 receptor binding sites as revealed by labelling with ^{125}I-SCH 23982 and show an increase in binding site density from 215 ± 62 fmol/mg protein to 701 ± 135 fmol/mg protein following treatment with a differentiating agent, *n*-butyrate, at a concentration of 1 mM for 6 days (71). In contrast, the parent neuroblastoma line in which D_1 receptor binding is approximately 35% of the hybrid cell line, is unaffected by treatment with *n*-butyrate. In addition, 10^{-4} M dopamine in the presence of 10^{-7} M haloperidol increases adenylate cyclase activity in the hybrid cell line by 56% (71). The cells obtained from the striatal fusions appear therefore to have potential for examination of developmental regulation of dopamine receptor–effector systems and as a possible source of receptor-enriched cells for the cloning of the mouse dopamine receptors, themselves.

5. Conclusions and future directions

The present chapter has discussed the technical considerations that are important for generating CNS clonal cell lines using the somatic cell fusion approach. We have also provided examples of how this technique has been employed to establish stable cell lines from the septal and hippocampal regions as well as from the mesencephalon and striatum. A major advantage of this approach is that the cell fusion technique does not appear to be limited to primary cell populations that are still mitotically active. For example, the generation of post-natal septal cell lines that exhibit all of the biosynthetic machinery to synthesize and release ACh strongly suggests that the 'immortalization' event involved a primary cell that was already post-mitotic and committed to a particular differentiation pathway. The ability to immortalize a neuron at this stage in development provides important tools for investigating the cellular and molecular mechanisms that mediate the formation of specific neural circuits. The fact that the dopaminergic cell line, MN9D, can distinguish between target and non-target cells is particularly intriguing with regard to obtaining cell populations that will allow for detailed cellular and molecular studies of neuronal differentiation and cell recognition.

Every methodology and experimental approach has its limitations. In general, it is important to remember that all cell lines, irrespective of the manner in which they were generated, represent 'abstractions' of the primary neurons they are used to study. It is therefore essential that any observations made employing cell lines be tested in primary cells in culture and ultimately in the whole organism. However, cell lines represent an unlimited source of material for analyses that cannot normally be performed in either primary culture systems or in the intact brain due to a limited source of tissue. The particular disadvantage of the fusion approach described above is that the growth characteristics of the hybrid cell lines are more difficult to control than many retrovirally transduced cell lines. This drawback probably relates to the overtly malignant nature of the neuroblastoma cell line that is employed as the fusion partner. It is not clear what oncogenes or combination thereof are responsible for maintaining cell division. Another disadvantage is that the use of a peripheral neural tumour cell line, the N18TG2 neuroblastoma, introduces properties of a cell type that normally does not participate in the development of specific central neuronal pathways. Although the hybrid MN9D still exhibits cell–cell recognition properties specific to the nigrostriatal pathway, it is not clear at this point whether the influence of the fusion partner has limited the scope of these recognition properties. In fact, recent work suggests that cells of a particular lineage express regulatory proteins which repress the transcription of genes specific to other lineages (72).

A major challenge for the future is to develop technologies that will overcome these limitations and improve upon the strategies that are currently available to immortalize neural cells. One possibility is to utilize the information and technology developed from both retroviral mediated oncogene transduction and cell fusion approaches (*Figure 8*). It is possible to establish progenitor cell lines from specific brain regions using the former approach. The ability to engineer expression of a particular oncogene in a primary cell allows for the potential of greater control over the growth characteristics of the resulting cell line. For example, the use of the temperature-sensitive mutant of the immortalizing SV40 large T antigen provides a potentially precise handle on cell division (8, 9). This antigen is stable at 33°C and will promote cell division. At 37–39°C the antigen is inactive and cells are no longer stimulated to divide. Using a retroviral construct containing this gene, one can develop and characterize a series of progenitor cell lines from a particular brain region of interest (73–76). As with other cell lines that have been developed with this technique, one would not expect such progenitors to differentiate to any great extent along a neural or glial pathway. However, such cells, should conserve the lineage features that are unique to the brain region from which they are derived. It should then be possible to select and mutate a particular progenitor cell line of interest so that it can be selected against in culture. The HPRT gene is particularly advantageous because it is X-linked and therefore requires only one mutational event. Chemical mutagenesis using agents such as ethane methyl sulfonate are quite efficient in generating HPRT-deficient mutants. The resultant HPRT-deficient cell line can then be employed as a fusion partner for primary cells from the same brain region, but at a later stage of development. This approach would eliminate the use of the neuroblastoma, would provide a growth promoting gene that is better defined and more amenable to control, and would allow for

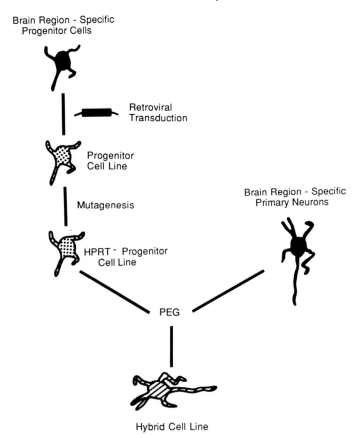

Figure 8. Schematic illustration of transduction/fusion strategy for generating CNS clonal cell lines. Embryonic brain cells (*black* = short processes) are infected with retroviruses containing an 'immortalizing' oncogene resulting in a 'progenitor' cell line (*stippled*). The progenitor cell line is subjected to chemical mutagenesis so that the HPRT gene is functionally eliminated. The HPRT⁻ progenitor cell line is then fused using PEG to mature primary neurons (*black* = long processes) from the same brain region, resulting in a hybrid cell line (*cross-hatched*) that is amenable to growth control, because of the oncogene transduction technique, and which preserves lineage features inherent in the brain region of interest.

the preservation of region-associated cell lineage. The resulting hybrid cell lines should express more highly differentiated phenotypes because of the potential of fusing the progenitor line to more mature primary neurons. Although the feasibility of such a strategy remains to be determined, it is clear at this time that there is a resurgence of interest in the generation and use of clonal cell lines from the CNS in a variety of studies. The scope of the major developmental neurobiological problems that remain to be answered will clearly require the availability of homogeneous cell populations that can be grown in unlimited quantities. The techniques and approaches discussed in this chapter and others represent a report of partial progress which is likely to experience considerable growth in the very near future.

Acknowledgements

We would like to thank Dr Lisa Won and Dr Hyung Choi for their critical comments in preparing this manuscript, and Mr Steve Price for his excellent technical assistance. This work was supported in part by RO1 NS-25787 and the Alzheimer's Disease and Related Disorders Society (BHW), RO1 MH28942 (AH), and grants to BHW and AH from the Brain Research Foundation, an affiliate of the University of Chicago.

References

1. Purves, D. (1988). *Body and brain: a trophic theory of neural connections.* Harvard University Press, Cambridge, Massachusetts.
2. Bottenstein, J. E. (1981). In *Functionally differentiated cell lines* (ed. G. Sato), pp. 155–84. Alan R. Liss, New York.
3. Schubert, D. (1984). In *Developmental biology of cultured nerve, muscle, and glia* (ed. D. Schubert), pp. 1–25. John Wiley, New York.
4. Banker, G. and Goslin, K. (1991). In *Culturing nerve cells* (ed. G. Banker and K. Goslin), pp. 11–39. MIT Press, Cambridge, Massachusetts.
5. Tishler, A. S. and Greene, L. A. (1975). *Nature,* **258,** 341.
6. Greene, L. A., Sobeih, M. M., and Teng, K. K. (1991). In *Culturing nerve cells* (ed. G. Banker and K. Goslin), pp. 207–26. MIT Press, Cambridge, Massachusetts.
7. Ronnett, G. V., Hester, L. D., Nye, J. S., Connors, K., and Snyder, S. H. (1990). *Science,* **248,** 603.
8. Lendahl, U. and McKay, R. D. G. (1990). *Trends Neurosci.,* **13,** 132.
9. Cepko, C. L. (1989). *Ann. Rev. Neurosci.,* **12,** 47.
10. Paige, C. J. and Wu, G. E. (1989). *FASEB J.,* **3,** 1818.
11. Alt, F., Blackwell, K., and Yancopoulos, G. D. (1987). *Science,* **238,** 1079.
12. Hammond, D. N., Wainer, B. H., Tonsgard, J. H., and Heller, A. (1986). *Science,* **234,** 1237.
13. Shay, J. W. (1982). *Techniques in somatic cell genetics.* Plenum Press, New York.
14. Kohler, G. and Milstein, C. (1975). *Nature,* **256,** 495.
15. Greene, L. A., Shain, W., Chalazonitis, A., et al. (1975). *Proc. Natl. Acad. Sci. USA,* **72,** 4923,
16. Platika, D., Boulos, M. H., Baizer, L., and Fishman, M. C. (1985). *Proc. Natl. Acad. Sci. USA,* **82,** 3499.
17. Hammond, D. N., Lee, H. J., Tonsgard, J. H., and Wainer, B. H. (1990). *Brain Res.,* **512,** 190.
18. Lee, H. J., Hammond, D. N., Large, T. H., and Wainer, B. H. (1990). *Devel. Brain Res.,* **52,** 219.
19. Lee, H. J., Hammond, D. N., Large, T. H., et al. (1990). *J. Neurosci.,* **10,** 1779.
20. Creps, E. S. (1974). *J. Comp. Neurol.,* **157,** 161.

21. Bayer, S. A. (1979). *J. Comp. Neurol.*, **183,** 89.
22. Semba, K. and Fibiger, H. C. (1988). *J. Comp. Neurol.*, **269,** 87.
23. Hemmendinger, L. M., Garber, B. B., Hoffmann, P. C., and Heller, A. (1981). *Brain Res.*, **222,** 417.
24. Gruneberg, H. (1943). *J. Hered.*, **34,** 89.
25. Hsiang, J., Wainer, B. H., Shalaby, I. A., Hoffmann, P. C., Heller, A., and Heller, B. R. (1987). *Neuroscience,* **21,** 333.
26. Morrow, J. (1982). *Techniques in somatic cell genetics* (ed. J. W. Shay), pp. 1–9, Plenum Press, New York and London.
27. Choy, W. N., Gopalakrishnan, T. V., and Littlefield, J. W. (1982). In *Techniques for using HAT selection in somatic cell genetics* (ed. J. W. Shay), pp. 11–21. Plenum Press, New York and London.
28. Killary, A. and Fournier, R. (1984). *Cell,* **38,** 523.
29. Augusti-Tocco, G. and Sato, G. (1969). *Proc. Natl. Acad. Sci. USA,* **64,** 311.
30. Lalley, P. A., Minna, J. D., and Franke, U. (1978). *Nature,* **274,** 160.
31. Minna, J. D., Marshall, T. H., and Shaffer-Berman, P. V. (1975). *Som. Cell Genet.,* **1,** 355.
32. Fournier, R. E. K. (1981). *Proc. Natl. Acad. Sci. USA,* **78,** 6349.
33. Mercer, W. E. and Baserga, R. (1982). In *Techniques in somatic cell genetics* (ed. J. W. Shay), pp. 23–24. Plenum Press, New York and London.
34. Choi, H. K., Won, L. A., Kontur, P. J., et al. (1991). *Brain Res.,* **552,** 67.
35. Dev, V. G. and Tantravahi, R. (1982). In *Techniques in somatic cell genetics* (ed. J. W. Shay), pp. 493–511. Plenum Press, New York and London.
36. Nesbitt, M. N. and Francke, U. (1973). *Chromosoma,* **41,** 145.
37. O'Brien, S. J., Simonson, J. M., and Eichelberger, M. (1982). In *Techniques in somatic cell genetics* (ed. J. W. Shay), pp. 513–24. Plenum Press, New York and London.
38. Siciliano, M. J. and White, B. F. (1987). *Methods in enzymology* (ed.), Vol. 151, pp. 169–194. Academic Press, Orlando, Florida.
39. DeLorenzo, R. and Ruddle, F. (1969). *Biochem. Genet.,* **3,** 151.
40. Farber, R. A. (1973). *Genetics,* **74,** 521.
41. Nilsson, O. G., Shapiro, M. L., Gage, F. H., Olton, D. S., and Bjorklund, A. (1987). *Exp. Brain Res.,* **67,** 195.
42. Wainer, B. H., Levey, A. I., Rye, D. B., Mesulam, M.-M., and Mufson, E. J. (1985). *Neurosci. Lett.,* **54,** 45.
43. Freund, T. F. and Antal, M. (1988). *Nature,* **336,** 170.
44. Wainer, B. H. (1989). *Neurobiol. Aging,* **10,** 540.
45. Cotman, C. W., Matthews, D. A., Taylor, D., and Lynch, G. (1973). *Proc. Natl. Acad. Sci. USA,* **70,** 3473.
46. Gage, F. H., Armstrong, M., Williams, L. R., and Varon, S. (1988). *J. Comp. Neurol.,* **269,** 147.
47. Hefti, F., Hartikka, J., and Knusel, B. (1989). *Neurobiol. Aging,* **10,** 515.
48. Fischer, W., Wictorin, K., Bjorklund, A., Williams, L. R., Varon, S., and Gage, F. H. (1987). *Nature,* **329,** 65.
49. Rosenberg, M. B., Friedmann, T., Robertson, R. C., et al. (1988). *Science,* **242,** 1575.
50. Knusel, B., Winslow, J. W., Rosenthal, J. W., et al. (1991). *Proc. Natl. Acad. Sci. USA,* **88,** 961.

51. Alderson, R. F., Alterman, A. L., Barde, Y.-A., and Lindsay, R. M. (1990). *Neuron,* **5,** 297.
52. Lee, H. J., Elliot, G. J., Hammond, D. N., Lee, V. M.-Y., and Wainer, B. H. (1991). *Brain Res.,* **558,** 1.
53. Carden, M. J., Trojanowski, J. Q., Schlaepfer, W. W., and Lee, V. M.-Y. (1987). *J. Neurosci.,* **7,** 3489.
54. Blusztajn, J. K. and Venturini, A. (1990). *Soc. Neurosci. Abs.,* **16,** 199.
55. Blusztajn, J. K., Venturini, A., Lee, H. J., and Wainer, B. H. (1991). I *FASEB J.,* **5,** A456.
56. Hsiang, J., Heller, A., Hoffmann, P. C., Mobley, W. C., and Wainer, B. H. (1989). *Neuroscience,* **29,** 209.
57. Bothwell, M. (1991). *Cell,* **65,** 915.
58. Kamegai, M., Niijima, K., Kunishita, T., *et al.* (1990). *Neuron,* **4,** 429.
59. Kamegai, M., Konishi, Y., and Tabira, T. (1990). *Brain Res.,* **532,** 323.
60. Wainer, B. H., Lee, H. J., Roback, J. D., and Hammond, D. N. (1990). *The basal forebrain: anatomy to function* (ed. P. W. Kalivas and T. C. Napier). Plenum Press, New York.
61. Hohn, A., Leibrock, J., Bailey, K., and Barde, Y.-A. (1990). *Nature,* **344,** 339.
62. Roback, J. D., Palfrey, H. C., and Wainer, B. H. (1991). *Curr. Topics Devel. Biol.* (In press).
63. Hornykiewicz, O. (1978). *Neuroscience,* **3,** 773.
64. Hornykiewicz, O. (1973). *Fed. Proc.,* **32,** 183.
65. Hyman, C., Hofer, M., Barde, Y.-A., *et al.* (1991). *Nature,* **350,** 230.
66. Porcher, W. and Heller, A. (1972). *J. Neurochem.,* **19,** 1917.
67. Joh, T. H., Baetge, M. E., Ross, M. E., and Reis, D. J. (1983). *Cold Spring Harbor Symp. Quant. Biol.,* Vol. 48, pp. 327–35. Cold Spring Harbor, NY.
68. Langston, J. W. and Irwin, I. (1986). *Clin. Neuropharmacol.,* **9,** 485.
69. Choi, H. K., Won, L., Hoffmann, P. C., and Heller, A. (1991). *Soc. Neurosci. Abs.,* **17,** 1275.
70. Choi, H. K., Won, L. A., and Heller, A. (1990). *Soc. Neurosci. Abs.,* **16,** 996.
71. Wainwright, M. W., Perry, B. D., Kontur, P., and Heller, A. (1990). *Soc. Neurosci. Abs.,* **16,** 646.
72. Schafer, B. W., Blakely, B. T., Darlington, G. J., and Blau, H. M. (1990). *Nature,* **344,** 454.
73. Eves, E. M., Lee, H. J., Tucker, M. S., Rosner, M. R., and Wainer, B. H. (1990). *Soc. Neurosci. Abs.,* **16,** 1149.
74. Tucker, M. S., Eves, E. M., Hou, X. Y., Wainer, B. H., and Rosner, M. R. (1990). E *Soc. Neurosci. Abs.,* **16,** 1149.
75. Tucker, M. S., Eves, E. M., Hou, X. Y., Wainer, B. H., and Rosner, M. R. (1991). *FASEB J.,* **5,** A1622.
76. Eves, E. M., Kwon, J., and Wainer, B. H. (1991). *Soc. Neurosci. Abs.,* **17,** 37.
77. Wainer, B. H., Lee, H. J., Roback, J. D., and Hammond, D. N. (1989). In *Novel approaches to the treatment of Alzheimer's disease* (ed. E. M. Meyer and J. Simpkins), pp. 71–94. Plenum Press, New York.

2

Generation of neural cell lines by transfer of viral oncogenes

PIERRE ROUGET, MARC LE BERT, ISABELLE BORDE,
and CLAUDINE EVRARD

1. Introduction

Many aspects of the regulation of expression of genes involved in neural differentiation and cellular interactions within the nervous system remain to be elucidated. This is largely due to the difficulty of establishing phenotypically untransformed permanent neural cell lines that retain most of the functional properties of their normally differentiated cell counterparts (1). Most established neural cell lines are transformed and have lost some aspects of normal regulation of cell division and differentiation. Even though presenting transformed or tumoral characteristics, however, some of these cell lines have provided useful models of neuronal function. Such lines have been obtained from spontaneous or chemically induced tumours (2, 3), or by transformation with oncogenic viruses (4, 5). Other cell lines have been derived from embryonal carcinoma (ref. 6 and Chapter 4) or from clones proliferating spontaneously in long-term primary cultures (7).

Alternative possibilities for establishing cell lines have been provided by the discovery of co-operation between polyoma virus oncogenes for the transformation of murine fibroblasts (8), shortly followed by similar findings with adenovirus and cellular-derived oncogenes (9–10). These observations have led to the definition of several molecular steps involved in the tumorigenic conversion of fibroblasts. The first of these steps, usually named immortalization, confers an unlimited growth potential on the cells, without leading to a transformed or tumoral phenotype. After transfection with the polyoma large T gene (8), or the adenovirus E1A region (9), or the viral or rearranged cellular *myc* gene (10), cells retain contact-inhibition of proliferation, do not show anchorage-independent growth and do not become tumorigenic (1, 8). In the case of the transfer of the Simian Virus 40 (SV40) large T gene the situation is somewhat different in that SV40 large T antigen carries both immortalizing and transforming functions (11); however, it has been shown that the transforming potential of large T is reduced when inserted in a retroviral vector (12). Taken together with the development of gene transfer methods, the above-mentioned observations have led to the establishment of different types of neural cell lines during the last few years. Some of them, corresponding to astroglial cells or to oligodendrocyte-type 2 astrocyte precursors, have been obtained by transfection of brain cell cultures with polyoma virus large T or adenovirus E1A sequences (1, 13) together with a *Neo^r* selection gene that has been co-introduced into the cells, either on the same vector as the oncogene, or by co-transfection. The use of recombinant retroviral vectors (14) for transferring oncogenes, has led to the generation of bipotential neuroglial precursor cell lines from different areas of the central nervous system (15–17), and also to the immortalization of sympatho-adrenal progenitors (18). Another approach to generating neuronal cell lines, based on the generation of hybrid cell lines (19), is discussed in detail in Chapter 1. A further approach for generating either immortalized or transformed neural cell lines has been based on the establishment in culture of clones isolated from non-tumoural brain regions of transgenic mice carrying the polyoma large T gene (20), or of cells derived from tumours that have been induced by a targeted expression of the SV40 large T gene (21, 22). In this chapter, methods and techniques for generating neural cell lines which are based on defined oncogene transfer or transgenesis are described, as are points concerning the analysis of the growth properties of such lines and methods for their characterization.

2. Preparation of cell cultures for transfection

2.1 Transient expression vs. integration

Although transient expression can readily provide information on the capacity of a vector to express a reporter gene and may give some indications as to the possible efficiency of a similar construct with an oncogene instead of a reporter gene, the extrapolation must be made with caution. For instance, the classical CAT-assay (chloramphenicol acetyl transferase-assay) (23) is carried out by measuring the enzyme activity in extracts from a whole cell population (see Chapter 8) and can reflect the expression of the CAT gene in a small subpopulation particularly sensitive to gene transfer or/and synthesizing trans-acting factors especially favourable to the regulatory region associated with the gene. It should be noted that a proportion of cells expressing the CAT gene, even less than 1% of the cell population, can lead to significant CAT activity which depends on the number of vector molecules transferred into the cells and on promoter efficiency. However, the *in situ* detection of beta-galactosidase activity due to *Lac Z* expression (24) does allow the determination of the proportion of the cells expressing the gene.

Immortalization requires the integration of the transferred oncogene into the genome and therefore, susceptible cells are those that have retained the potential for at least one division cycle. It should be kept in mind that, even when starting from dividing cells, the values that have been reported so far for the frequency of immortalization or transformation of neural cells, by means of oncogene transfection or transduction, ranged from 10^{-6} to 10^{-4} (1, 15, 17). For efficient transformation it is critical that prior to and up to 48 hours after transfection, the cultures do not reach cell densities beyond those indicated in *Protocols 1–3*.

2.2 Choice of developmental stage

Apart from the requirement for a round of cell division, the choice of developmental stage depends on the differentiated state and functional properties of the cells to be immortalized or transformed. This means that various gradients of development (rostro-caudal, dorso-ventral, latero-medial) must also be taken into consideration. As an example, the appearance of dopaminergic neurons in ventral mesencephalon occurs between embryonic day 12 and 15 (E12–E15), in the rat. This means that these cells cease to proliferate at this time, as shown by [^{3}H]-thymidine incorporation. This onset of neuronal maturation is rapidly followed by neurotransmitter synthesis: tyrosine hydroxylase, which is the first enzyme involved in dopamine synthesis, becomes detectable as soon as E12.5. In a more anterior brain structure, such as the striatum which receives afferent dopaminergic fibres from the substantia nigra, neuronal maturation is delayed by about 2 days, as compared with the mesencephalon.

In a defined region of the nervous system, the timing of neural differentiation depends on a complex set of factors including the cell type, its position in gradients of mitotic activity as well as its phylogenetic origin. For example in mouse cerebellar cortex, even though some overlap occurs, the Purkinje cells are formed first (E11–E14), then the Golgi type-II cells followed by basket cells and stellate cells, and finally by the granule cells (E17–P15). Note that some granule cell precursors are still dividing post-natally. Moreover, the onset of neurogenesis is dependent on the animal species: for instance, Purkinje cell maturation occurs between E3–E6 in chick embryo, E11–E14 in mouse, and E14–E17 in rat.

2.3 Cell culture conditions before transfection or transduction

In vivo, cell differentiation and proliferation are dependent on the expression of a genetic programme. This expression is itself modulated by cell–cell interactions. In cells in culture, survival, proliferation, and differentiation potential are influenced by several parameters including the developmental stage of the starting material, the amount of material at seeding, the composition of the culture medium, the timing of medium changes, and the coating of Petri dishes.

2.3.1 Typical media

The classical basal media for growing either primary neural cell cultures or immortalized cell lines are Dulbecco modified Eagle's medium (DMEM) with a high glucose concentration (4.5 g/litre) and DMEM-F12 1/1 (v/v). These media are generally supplemented with 0.11 g/litre sodium pyruvate, 50 IU/ml penicillin, 0.1 mg/ml streptomycin, supplemented either with fetal calf serum (FCS) (standard medium) or with newborn calf serum, depending on the cell type to be grown (see *Protocols 1–3*).

Some chemically defined media have been adapted to select different cell subpopulations for survival and proliferation. For example, media derived from the Bottenstein–Sato N2 medium (25) with minor modifications are commonly used for growing neuronal cells. A typical composition is: 5 μg/ml insulin, 100 μg/ml transferrin, 100 μM putrescin, 20 nM progesterone, 30 nM selenium. Defined media are discussed in detail in Chapter 3.

2.3.2 Preparation of the cells and growth conditions

A variety of protocols are presently available for cell cultures enriched in neuronal, astroglial, and oligodendroglial cells. Depending on the cell type to be grown, the tissue culture dishes should be coated with poly-ornithine, with poly-ornithine plus laminin, or with gelatin. Three examples are described for the above-mentioned cell types (see *Protocols 1–3*).

Protocol 1. Primary cultures of neurons

1. Preparation of tissue culture dishes
    - poly-ornithine coating: incubate the dish with 1.5 μg/ml D-L-poly-ornithine (M_r 40 000, Sigma) for 30 min at room temperature (RT). Wash the dish once with phosphate-buffered saline (PBS), and incubate with PBS supplemented with 10% FCS for 2 h at 37°C. Rinse the dishes twice more with PBS.
    - Laminin coating: coat the dishes with D-L-poly-ornithine as described above and then incubate them overnight with 1 μg/ml laminin (Sigma) in DMEM at 37°C.
    - Gelatin coating: cover the dishes with a solution of 250 μg/ml gelatin (Sigma) in PBS at RT. Wash the dishes once with PBS.
2. Dissect the region of interest in DMEM-F12 1/1 (v/v), without serum, choosing an embryonic stage when at least some neuronal precursors are still dividing (for example, mouse E-14 striatum).
3. Gently dissociate the tissue by trituration in the same medium (1 ml for 10–20 striata), using a Pasteur pipette (with fire-polished tip), until the cells are suspended.

4. Centrifuge the suspension at 500 g for 5 min. Discard the supernatant and resuspend the pellet in N2 defined medium, in order to prevent the proliferation of astroblasts.
5. Plate the cells at $10^4 - 5 \times 10^4$ cells per cm^2, on poly-ornithine-coated dishes containing N2 defined medium. Incubate the cultures at 37°C, in a humidified 5% CO_2 incubator. Transfection or transduction should be carried out within the first 12 h post-seeding, whilst some neuroblasts are still dividing.

Protocol 2. Primary and secondary cultures of astrocytes

1. Prepare tissue culture dishes as described in *Protocol 1*.
2. Dissect the region of interest, choosing an embryonic stage when astrocytes are still dividing, whereas the majority of neuronal precursors have ceased proliferation (for example, E15–16 mouse striatum).
3. Dissociate the cells in DMEM and centrifuge them as described in *Protocol 1*, steps 2 and 3. Resuspend the cells in standard medium (see Section 2.3.1).
4. Seed the cells at a density of 5×10^4 cells per cm^2 on poly-ornithine-coated dishes containing standard medium.
5. Grow the cells at 37°C, for 1 to 3 weeks, in order to generate astroglial monolayers. Change the medium twice a week.
6. Detach the cells by trypsinization: rinse the monolayers twice with 0.02% EDTA in calcium- and magnesium-free PBS. Incubate with 0.125% trypsin–0.01% EDTA in calcium- and magnesium-free PBS for 20 min at 37°C. Stop the trypsinization by adding FCS to a final concentration of 10%, collect the cells by centrifugation (500 g for 5 min). Resuspend the pelleted cells in standard medium.
7. Plate 10^4 cells per cm^2 on gelatin-coated dishes, so that about 5×10^3 cells will be attached 24 h later. At this time, change the medium and carry out the transfection.

Protocol 3. Cultures enriched in oligodendroglial cells

1. Prepare tissue culture dishes as described in *Protocol 1*.
2. Dissect the desired brain region at a developmental stage when oligodendroglial cells are still dividing, whilst the proliferation of astrocytes is declining (for instance, brain hemispheres from newborn rat). Carefully remove the meninges.
3. Stretch a Nitex nylon filter (10 × 10 cm, 82 μm mesh) over a 60-mm dish

Protocol 3. *Continued*

containing 5 ml DMEM supplemented with 10% newborn calf serum. Place 5–6 hemispheres in the centre of the nylon filter and immerse them in the medium. Gently dissociate the tissue through the mesh, using a glass rod.
4. Pour the cell suspension into a tube containing 20 ml of the same medium.
5. Change the nylon filter and repeat the procedure with another pool of 5–6 hemispheres. Collect the cell suspensions in the same tube.
6. Seed the cells at a density of 1.5×10^5 cells per cm^2, on poly-ornithine coated dishes containing 5 ml of the same medium. Change the medium on days 5 and 8 post-seeding. This procedure results in a mixed glial culture containing up to 50% oligodendrocyte precursor cells after 7–9 days in culture. Transfection or transduction should be carried out at this time.

3. Transformation with oncogenic DNA viruses

Infection with oncogenic DNA viruses may give rise to a lytic cycle in some animal species resulting in virus multiplication and the death of permissive cells. In semi-permissive or non-permissive cells, the infection is abortive and may lead to the integration of viral DNA into the cell genome and to the oncogenic transformation of the cells. Thus polyoma virus can replicate and be propagated in mouse cells, whereas in rat cells, it gives rise to cell transformation. For SV40, monkey kidney cells are permissive and thus lytically infected, while mouse cells are not permissive for viral replication, but are susceptible to transformation.

Analysis of the genome of these viruses and transfection studies with cDNAs encoding polyoma virus early proteins, have led to the conclusion that transformation occurs through several molecular steps. However, when either the whole genome, or the virions themselves are used for transfection or infection respectively, non-permissive cells acquire a fully transformed phenotype (see Section 5.2). Nevertheless, owing to the simplicity of the infection, SV40 has been used to generate permanent cell lines. Although transformed, some of them have given useful insights into cell differentiation and cell–cell interactions (4, 5).

Protocol 4. Preparation of SV40 stocks and generation of SV40 transformed cell lines

1. Seed MA 134 or CV1 cells (ATCC) in 175-cm^2 flasks containing 20 ml DMEM medium supplemented with 10% FCS.
2. When the cells reach subconfluence, discard the medium and infect the cells with 1.5 ml of a viral suspension in DMEM, at a multiplicity of 0.1–0.2 plaque forming units (p.f.u.) per cell.

3. Incubate the cells with virus for 1–2 h at 37 °C (5% CO_2). Then add 20 ml DMEM containing 42 mM Hepes (pH 7.2), supplemented with 2% FCS. Replace the medium on days 3 and 6 post-infection. The cells will lyse within the following two days.

4. Verify that lysis is complete, before freezing the flasks at −70 °C. After three freeze–thaw cycles to disrupt the cells, centrifuge the suspension at 12 000 g for 10 min and collect the supernatant. Store the supernatant containing virus in aliquots at −70 °C.

5. To determine the virus titre, grow CV1 cells in 60-mm dishes containing DMEM with 10% FCS.

6. Prepare serial dilutions (10^{-5} to 10^{-8}) of the viral stock and infect CV1 monolayers with 0.2 ml of these dilutions. Incubate the cells for 2 h at 37 °C (5% CO_2).

7. Replace the medium with 5 ml of semi-solid medium (growth medium containing 0.5% gelose) and add a further 2 ml of this medium at 3-day intervals.

8. Calculate the virus titre by counting the plaques of lysed cells which appear after 8–12 days. The plaques can be scored directly by eye. Alternatively, after removal of medium, fix the cell monolayer with 4% paraformaldehyde in PBS for 10 min, wash the cells with PBS, and stain the monolayer with 5% v/v Giemsa stain in PBS.

9. Generate cell lines transformed with SV40 using primary or secondary cultures prepared as described in *Protocols 1–3*. Remove the medium, add the viral suspension (0.2 ml at 10^8–10^9 p.f.u./ml per 60-mm dish) and incubate the cells with virus for 1–2 h at 37 °C before washing and adding the appropriate growth medium. Directly dissociated cells can also be treated with the same infection procedure.

---

4. Transfection of cells with oncogenes

4.1 Oncogenes and vectors for transfection-mediated immortalization

4.1.1 Typical vectors

The first immortalizing constructs were eukaryotic expression vectors carrying oncogenes controlled by their own promoter and enhancer sequences. They carried either the polyoma large T gene (8), the adenovirus E1A sequence (9), a rearranged c-*myc* from the mouse plasmacytoma MOPC315 or v-*myc* from the avian myelocytomatosis MC29 (10).

In order to provide a convenient selection of the cells that have stably integrated the genes, a drug-resistance gene should be introduced simultaneously with the oncogene, either on the same plasmid, or by co-transfection

Generation of neural cell lines by transfer of viral oncogenes

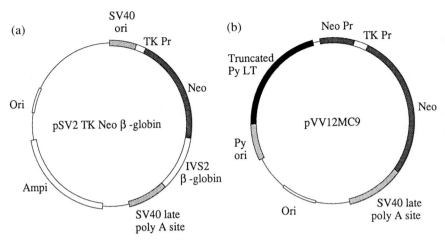

Figure 1. (a) Illustrates the organization of the selection vector, pSV2 TK Neo β-globin (29), and (b) shows the vector pVV12MC9, carrying a truncated form of the polyoma virus large T gene and the *Neo'* gene (30).

(1, 13). The most commonly used selectable genes are *Neo'*, *Hph'*, and *Mtx'* which confer resistance to G418 (26), hygromycine (27), and methotrexate (28), respectively. In typical constructs, they have been associated with the SV40 control region that itself can also be fused with the thymidine kinase promoter (29) or with a mixed prokaryotic–eukaryotic promoter (30). Two schematic drawings of a *Neo'*-selection vector, and of a vector carrying both an immortalizing sequence and a *Neo'*gene are presented in *Figure 1a* and *b* respectively.

Note that when the oncogene and the drug-resistance gene are carried by different vectors and are introduced by co-transfection, the ratio between the selectable marker and the oncogene vectors should be 1:10 to ensure that all drug-resistant cells have also received the oncogene.

4.1.2 Inducible promoters and conditional mutants

One method of controlling the expression of the introduced oncogene is to associate it with inducible promoters; for example, those from heat-shock protein genes, metallothionein, or growth hormone genes, or sequences derived from the long terminal repeat (LTR) of the mouse mammary tumour virus (MMTV). However, these systems present several limitations: the heat shock response is transient, and the metallothionein promoter leads to some background level of expression in the absence of inducers. Moreover, the induction stimuli often alter the expression of other cellular genes: for instance, astroglial cell lines that are glucocorticoid dependent for proliferation have been obtained with constructs where adenovirus gene was controlled by sequences from the MMTV-LTR (31). Even though these lines can be used for studying the control of cell division, they have been somewhat disappointing for studying differentiation, probably due to the fact that glucocorticoids also exert an effect on glial differentiation as shown on primary cultures (32, 33).

Thermosensitive oncogene mutants such as tsa mutants from SV40 have also been used, in spite of their relative leakiness at non-permissive temperatures. Constructs with the SV40 tsa large T gene inserted in retroviral vectors (see Section 5) have lead to neural precursor cell lines able to stop proliferation and to differentiate when shifted to the non-permissive temperature (16). Targeted expression vectors carrying oncogenes have also been introduced into transgenic mice and these are briefly discussed at the end of the chapter.

4.2 Calcium phosphate precipitation

The most widely used method for introducing foreign DNA into eukaryotic cells is based on the simultaneous precipitation of DNA when insoluble calcium phosphate is formed by mixing $CaCl_2$ with a sodium phosphate solution. Athough the mechanism of entry of DNA into the cells remains unclear, it is generally assumed that the precipitate enters by endocytosis, and then is transferred to the nucleus. The majority of the reported procedures have been derived from the initially described experimental conditions (34), with minor modifications. A typical procedure is described in *Protocol 5*.

The following stock solutions should be prepared, sterilized by filtration and stored at $-20\,°C$:

- 2 M $CaCl_2$ in water
- 2 × Hepes-buffered saline (2 × HBS): 40 mM Hepes, 280 mM NaCl, and 1.5 mM Na_2HPO_4. Adjust the pH to 7.1 with HCl.
- 2 × Bes-buffered saline (2 × BES): 50 mM Bes, 280 mM NaCl, and 1.5 mM Na_2PO_4

Adjust the pH to 6.95.

Protocol 5. Calcium phosphate precipitation

The following quantities are appropriate for cells growing in a 60-mm dish containing 5 ml medium.

1. Mix 5 to 10 μg of vector DNA with 31 μl of 2 M $CaCl_2$ and bring the final volume to 0.25 ml with H_2O.
2. Pour this solution dropwise into 0.25 ml of 2 × HBS[a] and incubate for 10 min at RT to allow the precipitate to develop.
3. Add the precipitate in the medium[b] and put the dish back into the incubator immediately, in order to ensure that the pH of the medium does not change.

Protocol 5. *Continued*

4. Incubate the cells for 15 h at 37°C, in a 5% CO_2 incubator.
5. Wash the cells twice with serum-free medium and then[c] add 5 ml standard medium supplemented with 10% FCS.

[a] An alternative procedure that may be more efficient for gene transfer (35), provided that the viability of the cells is unaffected, is to use Bes instead of HBS and incubate the cells under 3% CO_2.

[b] When the precipitate is not toxic for the cells, its direct addition on to the monolayer also improves the transfection efficiency: remove the medium, pour the precipitate dropwise on to the cells, incubate 20 min at RT, add 5 ml standard medium before continuing with steps 4 and 5.

[c] A glycerol-shock may be carried out at this time in order to increase the efficiency of transfection, although cell viability may also be affected. Wash the cells, add 0.5 ml of 15% glycerol in serum-free medium, for 30 sec to 3 min (depending on the cells). Wash the cells again before feeding the cells with the standard medium.

4.3 Electroporation

This method is based on the delivery of high voltage-electric pulses which induce transient pore-like structures allowing the DNA to enter the cells. As compared with the calcium phosphate precipitation procedure, electroporation generally allows a higher proportion of cells to be transfected, although with a lower vector copy number per cell. This can be verified in parallel transient assays with expression vectors carrying either the *CAT* gene or the *LacZ* gene: the first system allows gene expression to be easily quantified in cell extracts (23), while the second leads to a convenient determination of the proportion of the cells expressing the gene (24).

Even though it is possible to carry out electroporation on attached cells, a higher cell number can be treated with cells in suspension, either directly dissociated from dissected brain regions, or trypsinized from primary or secondary cultures.

4.3.1 Electroporation apparatus

Several electroporation systems are commercially available, which differ in the shape of the electrical pulse that is generated.

(a) Capacitor-discharge exponential decay-pulse (0.4–1.2 kV/cm) where the exponential decrease has a time-constant equal to the product of the capacitance and the sample resistance (36). By varying the capacitance, the duration of the pulse can be adjusted. This type of equipment can be obtained from BRL (Cellporator electroporation system) or Pharmacia (Isco 494).

(b) Square wave electric pulse where the field strength, duration, and frequency of pulses are simultaneously programmed (37).

4.3.2 Pulsing buffer

Different ionic compositions should be assayed to optimize the efficiency for different cell populations.

Two typical solutions are:

- 10 mM sodium phosphate buffer (pH 7.4), 75 mM NaCl, 1 mM KCl, 33 mM D-glucose, 92 mM saccharose: modified from (37)
- 0.7 mM Na_2HPO_4, 20 mM Hepes pH 7.05, 137 mM NaCl, 5 mM KCl, 6 mM D-glucose (38)

An additional related protocol is provided in Chapter 4.

Protocol 6. Electroporation

1. Prepare the cells by washing them once with the pulsing buffer and resuspending them at 3×10^6–10^7 cells per millimetre. Add 10–50 µg of linearized DNA to 1 ml of the cell suspension and incubate it at 4°C for 15 min.
2. Introduce the cell samples between the electrodes; two different systems are commonly used:
   - Sterile propylene electroporation chambers with two flat aluminium electrodes 0.4 cm apart are commercially available (BRL). They should be filled with 0.75 ml of the cell suspension.
   - Two stainless steel electrodes seated on the bottom of a dish with a volume of 0.1–0.2 ml between the electrodes
3. Apply the electric pulses: usually 1–5 pulses of 1–1.4 kV/cm and of 50–100 µsec, with 1–3 sec intervals between each pulse, are used.
4. Remove the cells and keep them in the electroporation buffer at RT for 10 min before addition of growth medium.
5. Check the viability of the cells 10 min after electroporation: mix 10 µl of the cell suspension with 10 µl trypan blue stain (0.16% trypan blue, 0.14 M NaCl) and count the viable refringent cells vs. the stained dead cells.

4.4 Lipofection

This method is based on the interactions of the plasma membrane with synthetic lipids carrying fatty acids associated with polycationic molecules (polyamines) that bind the DNA (39). These complexes form structures reminiscent of unilamellar liposomes (40) that are able to fuse with the membrane and to deliver DNA into the cells.

Protocol 7. Lipofection

The following quantities are appropriate for cells growing in a 60-mm dish:

1. Add 5–10 μg plasmid DNA to 1 ml serum-free medium.
2. Add 7.5–50 μl of a 1 mg/ml lipospermine (Transfectam, IBF) or lipofectin (Gibco-BRL) solution to 1 ml serum-free medium: use an amount of lipospermine in order to obtain a ratio lipospermine/DNA (w/w) ranging between 1.5 and 5.^{a}
3. Mix the two above-mentioned solutions and pour the mix on to the cells, previously washed twice with serum-free medium.
4. Maintain the cells for 1–6 h^a in an incubator set at 37°C, 5% CO_2.
5. Add FCS to a final concentration of 20%, incubate for a further hour, and then replace the medium with standard growth medium.

^{a} The amount of lipospermine and the incubation times depend on the toxicity of lipospermine with the particular cell type being utilized. This must be empirically determined.

5. Transduction of oncogenes with recombinant retroviral vectors

5.1 Recombinant retroviral vectors

Another approach to the generation of permanent cell lines is provided by the use of recombinant retroviral vectors carrying an oncogene. In recombinant retroviral vectors, most of the structural genes of the retrovirus have been deleted and replaced with heterologous genes. In order to obtain a viral particle able to transduce these genes, the recombinant genome of the vector must be complemented by the viral *gag*, *pol*, and *env* structural genes, in the form of complementary DNA, that has been integrated into the genome of helper (packaging) cell lines (see Section 5.2). The main advantage of recombinant retroviral vectors is that the transduction of the inserted heterologous genes is very efficient. Integration occurs rapidly and a single copy of recombinant provirus DNA is integrated per cell genome, even though this occurs at apparently random sites.

A limitation in the use of retroviral vectors is that, after reverse transcription, the integration of the recombinant retroviral vector DNA requires a round of host DNA synthesis. A further limitation is that appropriate receptors for the viral envelope glycoproteins must be expressed on the surface of the cell-type of interest.

5.2 Trans-complementation between the vector and helper sequences

As recombinant retroviral vectors do not carry the structural genes encoding the viral proteins necessary for the production of retroviral particles, these proteins must be encoded by another retroviral genome. For this purpose, helper cell lines, called packaging cell lines, have been derived by transfection of a retroviral genome (in the form of DNA) encoding all the viral structural proteins, but deleted of a sequence Ψ that carries the signal necessary for the viral genome to be packaged (see *Figure 2a* and *Figure 3a*). The viral *gag*, *pol*, and *env*, structural genes have been integrated into the genome of the helper cell line, but the absence of the packaging signal impedes the helper cell line from producing wild-type virions. In contrast, when recombinant retroviral vector DNA carrying the Ψ sequence is introduced into the packaging cell line, the recombinant genome is efficiently packaged, thus producing virions able to transduce the gene inserted in the vector (see *Figure 2b*). In terms of the infection process, these recombinant retroviral vector particles do not carry viral structural genes, and thus cannot produce new viral particles after infection. They behave as 'one-hit' viruses.

However, it is important to be aware of the possibility of recombination between vector and helper genomes, that could lead to the recovery of the Ψ sequence by the helper genome and thus to the production of replication-competent virus. Sensitive assays for detecting such viruses in recombinant retroviral stocks are available. A convenient method is to incubate NIH-3T3 cells with the supernatant of cells that have been infected with the recombinant retroviral particles (see *Protocol 8*). Alternatively, use the S^+L^- plaque assay that is described in detail elsewhere (41).

5.3 Typical recombinant retroviral vectors carrying an oncogene

Most of these vectors have been derived from the constructs initially designed by Cepko *et al.* (14). This group inserted different oncogenes into a backbone construct called pZIP-NeoSV(X)1. This type of vector, carrying the wild-type or tsa mutant SV40 large T gene, or else the v-*myc* gene, has been found to be suitable for the immortalization or the transformation of different neural precursor cells (16–18). A schematic illustration of the general organization of such vectors is outlined in *Figure 3b*. The vector pZIP-NeoSV(X)1 is a derivative of the Moloney murine leukemia virus (M.MuLV) (see *Figure 3a*). Between the M.MuLV long terminal repeats (LTR), the vector contains the Ψ packaging sequence, the donor (SD) and acceptor (SA) splice signals of the virus, whereas a *Neo^r* gene replaces the *env* viral gene and a *Bam* H1 cloning site allows heterologous genes to be inserted at the gag position. In these vectors, the transcription of the heterologous genes is initiated from the 5' LTR.

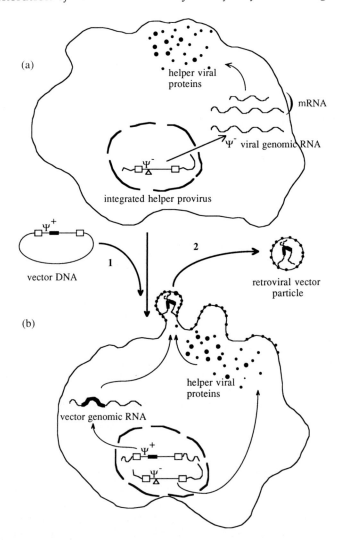

Figure 2. A helper packaging cell line is illustrated in (**a**): the helper proviral sequences express all the viral functional proteins, but the proviral DNA and the viral genomic RNA are deleted of the Ψ packaging-signal sequence, and thus the genomic RNA cannot be encapsulated. The *arrow* 1 indicates the introduction of a recombinant retroviral vector carrying the Ψ sequence and heterologous genes, but deleted of all the viral structural genes, into the packaging line. After vector integration, the cell line becomes able to produce both the viral proteins and the vector RNA that carries the Ψ sequence, and thus that can be packaged as outlined in (**b**). The *arrow* 2 shows the production of the recombinant retroviral particles which are able to transduce the genes carried by the vector.

(a) M.MuLV

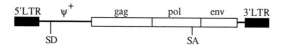

(b) pZIP-Neo SV(X)1

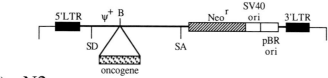

(c) pN2

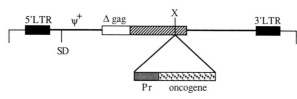

Figure 3. [a] Represents the linear map of the proviral DNA of M.MuLV, with the two LTR, the Ψ sequence and the viral structural genes *gag*, *pol*, *env* (SD and SA indicate the position of a splice donor and of a splice acceptor site). (b) and (c) illustrate two types of recombinant retroviral vectors: in the pZIP-NeoSV(X)1 vector (14) the oncogene can be inserted at the *Bam*H1 site (indicated by B), and can be expressed from the viral LTR. In the pN2 vector (42), the oncogene associated with a eukaryotic promoter (Pr) is inserted at the XhoI site (indicated by X).

In addition, such vectors contain the SV40 and pBR322 replication origin sequences and thus can also be used as shuttle vectors.

An alternative strategy is to insert an oncogene in the recombinant retroviral vector pN2 (42). In these constructs, the oncogene associated with a eucaryotic promoter can be inserted at the Xho 1 site which is close to the 3'-end of the *Neo^r* sequence (*Figure 3c*). Such a vector with the oncogene v-*src* has been reported to immortalize neural precursors which were blocked in their differentiation, presumably due to v-*src* expression (43). Other derivatives of this type of vector, carrying c-*myc* or N-*myc* have been used for immortalizing neuroepithelial or neural crest cells able to differentiate *in vitro* (15, 44).

5.4 Packaging cell lines

Several packaging cell lines are available. The Ψ-2 line (45) carries a M.MuLV provirus, where the Ψ sequence is deleted, but encoding structural proteins compatible with the infection of murine cells. The PA12 and PA317 lines lead to the production of amphotropic recombinant retroviral viruses able to infect cells from a large spectra of animal species (46). As these amphotropic retroviral particles can infect human cells, they must be used with the greatest caution, especially when the vector itself carries an oncogene.

In order to prevent possible recombination between vector and helper genome, some packaging cell lines have been derived by transferring helper genes by cotransfection with two constructs separately encoding the viral proteins (47).

5.5 Production of recombinant retrovirus stock

After transfection of packaging cells with an appropriate vector carrying an oncogene and a selectable marker gene, as indicated in *Protocol 5*, the cells that have integrated the vector should be selected and cloned as described in *Protocol 10*.

To produce recombinant retroviral stocks:

(a) Grow the producer cells in 60-mm dishes containing 4 ml of standard medium. When the cells become half-confluent, change the medium and return them to the incubator overnight. Take the supernatant and pass it through a 0.45 μm filter.

(b) Higher titre virus can be produced by removing the medium and adding 0.2 ml fresh medium to the monolayer every 2 h. Collect the medium and use it immediately to infect the cells. Repeat this procedure 2–3 times.

(c) If necessary, the viral supernatant can be concentrated by centrifugation at 25 000 r.p.m. in a Beckman SW28 rotor or equivalent. The pellet is then suspended in 1% of the original supernatant volume. It can be either used immediately or divided into aliquots and stored frozen in liquid nitrogen.

The titre of the recombinant virions should be determined as indicated in *Protocol 8* and the lines producing the highest titre should be used as a source of recombinant virus.

5.6 Determining the recombinant retrovirus titre

The titre of the virus stock (active virions/ml) is generally determined by infection of cultured fibroblast lines, the most commonly used being NIH-3T3 cells. The expression of proteins encoded by the vector is directly assayed, by checking the resistance to a drug (such as G418, hygromycine, methotrexate) or β-galactosidase activity or else immunoreactivity of the oncogene product. For example, to test the G418-resistance conferred by the expression of the *Neo^r* gene proceed as described in *Protocol 8*.

Protocol 8. Determination of recombinant retrovirus titre

1. The day before infection, seed 10^5 NIH-3T3 cells in a 60-mm dish containing standard medium.
2. Infect the cells as described in *Protocol 9*.
3. Twelve hours later, trypsinize the cells and plate them at low cell density (10^2–10^4 cells per 100-mm dish).
4. One day later, add 500 μg/ml G418 (Geneticine, Gibco). Change the selection medium every 3–4 days. The G418-resistant colonies will appear within 10 days.
5. Count the colonies. The titre will be equal to the number of colonies obtained per millilitre of viral suspension used for infection.

5.7 Retroviral vector-mediated transduction

After transfection of packaging cells with the appropriate vector carrying an oncogene and a selection gene, according to *Protocol 5*, infect the cells as described in *Protocol 9*.

Protocol 9. Infection with recombinant retroviral particles

1. Seed the chosen cells at densities depending on the cell type, as indicated in *Protocols 1–3* (last step).
2. Add 1 ml of the virus stock plus 4–8 μg polybrene (Sigma), a polycation that improves the adsorption of retroviral particles.
3. Incubate the cells at 37°C, 5% CO_2, for 2 h.
4. Remove the medium and replace it with growth medium.
5. Select the cells as described in *Protocol 10*.

The efficiency of transduction can be improved in a number of ways:

- High titre virus stock can be produced as described in Section 4.4.
- Alternatively, the cells to be infected can be seeded directly on to virus producing cells that have been rendered non-proliferative either by irradiation or by mitomycin C treatment, as follows:

    Irradiate the producer cells at 2800 rads and plate them at 6×10^3 cells per cm^2. As soon as they have attached, seed the cells to be infected directly on to these irradiated cells in standard medium containing 4 μg/ml polybrene. One day later, remove the cells by trituration and replate them on dishes previously coated as described in Section 2.3.2.

Generation of neural cell lines by transfer of viral oncogenes

For the mitomycin procedure, treat the subconfluent producer cells with 10 μg/ml mitomycin C (Sigma) for 2 h and wash three times with the culture medium. Seed the cells to be infected on to the producer cells as described above.

6. Selection and growth properties of cell lines

6.1 Selection and establishment of cell lines

When a drug-resistance gene has been introduced into the cells, either on the same vector, or by co-transfection, the cells that have integrated the vectors can be selected on the basis of their acquired resistance to the drug. This type of selection is needed when the cells to be immortalized may be overgrown by contaminating cells or when they have retained a proliferation potential initially sufficient to mask the effect of the immortalizing genes.

Even though the procedure presented in *Protocol 10* has been found to be efficient for different neural cell types, some modifications may be necessary for selecting particular neuronal subpopulations. Cell immortalization or transformation by means of oncogene transfer is a relatively new technique and so, for each particular cell population, preliminary experiments should be carried out, to determine the drug concentration necessary to kill untransfected cells.

Protocol 10. Selection of immortalized clones

1. After transfection or transduction (see *Protocols 5–7*), grow the cells for at least 24 h before adding the selection drug, in order to allow the expression of the drug-resistance gene to take effect.
2. Depending on the selection system, add the following concentrations of drugs:
   - 150–200 μg/ml G418
   - 50–100 μg/ml hygromycine
   - 0.1–1 μM methotrexate
3. Renew the selection medium every 3–4 days, for 4–6 weeks, until resistant colonies appear.
4. Replace the selection medium with the appropriate growth medium (see *Protocols 1–3*).
5. Trypsinize the colonies individually within cloning cylinders and transfer them into 15 mm-diameter wells.
6. Subclone the immortalized lines by limiting dilution and then expand them sufficiently to allow a frozen stock to be established.

Pierre Rouget, Marc Le Bert, Isabelle Borde, and Claudine Evrard

6.2 Growth properties of cells: immortalization vs. transformation

In order to select immortalized cell lines with properties as near as possible to those of their normal cell counterparts, it is of interest to check to what extent the cell lines have acquired a transformed phenotype associated with the ability to grow in the absence of anchorage, the capacity to form cell multilayers, their tumorigenicity in nude mice, and an aneuploid karyotype.

The anchorage dependence of cell growth, can be assayed by measuring growth in semi-solid medium as described in *Protocol 11*.

Protocol 11. Growth in soft agar

1. Prepare agar underlayers with 5 ml standard medium containing 0.5% agarose, 60-mm Petri dishes.
2. Prepare 2×10^4–3×10^5 cells in 3 ml of growth medium containing 0.3% low-melting-point agarose or 1.2% methocel.
3. Seed the cells on to the agarose underlayer and allow the cells to grow for 3–4 weeks (37°C, 5% CO_2) with regular changes of growth medium.
4. Examine the plates regularly by eye for evidence of colony formation characteristic of transformed cells.

To assay tumorigenicity, inject 2×10^6–2×10^7 cells subcutaneously between the ears of 5–10 nude mice. Screen for the presence of a palpable tumour mass after 1 and 2 months. It is vital that any work with animals conforms to legal guide-lines and is carried out by fully trained personnel. A further indicator of transformation is the presence of an abnormal number of chromosomes in the cell line of interest. *Protocol 12* describes how to determine the karyotype of cultured cells.

Protocol 12. Metaphase chromosome preparation

1. To accumulate the cells in metaphase, add colcemid (0.04 μg per millilitre of medium), to exponentially growing cells. Incubate the cells at 37°C for 2 h.
2. Collect the medium containing detached cells, trypsinize the cells that remain attached (*Protocol 2*), mix them with the medium and pellet the cells at 800 g for 10 min.
3. Resuspend the cell pellet in 10 ml 1% trisodium citrate (the volumes indicated in this protocol are for 2×10^6 cells). Incubate the cells in this hypotonic solution for 10–20 min at 37°C.
4. Add 1 ml of a Carnoy fixative solution (3:1 methanol–acetic acid, v/v),

Protocol 12. *Continued*

dropwise and with gentle mixing. The Carnoy solution should be prepared immediately beforehand and chilled to −20°C. Pellet the cells as indicated in step 2.

5. Add 8 ml of Carnoy solution dropwise, centrifuge the cells and resuspend the pellet in a further 1 ml of Carnoy solution.
6. To spread the chromosomes effectively, use microscope slides that have been previously washed with ethanol and kept at −20°C. Incline the slide slightly (about 30 degrees to the horizontal) and drop approximately 50 µl three times from a Pasteur pipette held 15 cm above the slide. Immediately pass the back of the slide briefly over a Bunsen flame. This temperature shock leads to the disruption of the cells and the spreading of the chromosomes over the slide, with little intermingling of the individual metaphases.
7. Stain the chromosomes with a solution of 10% (v/v) May-Grunwald Giemsa, 3% (v/v) methanol in 8 mM sodium citrate buffer (pH 5.75). Rinse with this citrate buffer, and air-dry the slides.

Corresponding control experiments can be carried out either with tumour cells, or with the examined immortalized cell line after supertransfection with transforming sequences, such as the polyoma middle T or the adenovirus E1B genes.

6.3 Status of the transferred oncogene and clonality of cell lines

In order to verify that the transferred oncogene has been integrated into the cell genome and to determine the number of integrated copies, genomic DNA should be digested with appropriate restriction enzymes (preferably leading to 0 or 1 cut per vector DNA), and then subjected to Southern blot analysis (48) with a probe specific for the oncogene. Such an analysis allows the clonality of the lines to be verified. After transfection or retroviral transduction, the integration of the vector occurs at random in the cell genome. Therefore a restriction pattern corresponding to a single integration site (even possibly with tandem repeats) implies that the cell line is clonal. This verification is particularly important when the cell line behaves as a bi- or multipotential precursor cell for several neural cell lineages.

6.4 Expression of the oncogene

The product of the oncogene can be detected with specific antibodies, by immunocytochemical assays, following a procedure similar to that described for the detection of internal antigens (see Section 8), or by immunoprecipitation or immunoblotting experiments. When looking for a correlation between oncogene expression and cell proliferation, it may also be useful to quantify the oncogene mRNA by Northern blot hybridization with specific probes (48) and to compare the actively dividing cells with the confluent and possibly growth-arrested cells, for this expression.

6.5 Precautions for avoiding secondary transformation events

It is very important to freeze cells from early passages, to prevent cells from acquiring transformed characteristics and losing their differentiated characteristics. As non-transformed cells are contact-inhibited at confluence, any actively dividing spontaneous transformant will be selected when the cells are maintained at confluence. It is therefore imperative to passage the cells at densities less than those at which contact inhibition occurs.

Establishing frozen stocks of cells:

- Wash the plates with EDTA 0.02% in calcium- and magnesium-free PBS.
- Dissociate the cells with 0.125% trypsin–0.01% EDTA in calcium- and magnesium-free PBS for 5–10 min at 37°C and stop the trypsin action by the addition of FCS (10% v/v final).
- Spin the suspension and resuspend the cells at a density of 2×10^6 cells per millilitre of freezing-medium, which comprises: dimethylsulfoxide (DMSO) 10% or glycerol 10%, FCS 20% in growth medium.

7. Characterization of cell lines: differentiation and functional properties

A convenient way to characterize cell lines is to analyse the expression of differentiation markers by immunocytochemical methods. Whenever possible, the results should be confirmed by immunoblotting analysis of the proteins and by Northern blot experiments with specific probes for studying RNA expression. These methods and the corresponding procedures are exhaustively documented elsewhere (48).

7.1 General considerations on the differentiation of permanent cell lines

Many permanent cell lines do not express any known differentiation markers, either spontaneously or even after attempts to induce them by various agents of differentiation. Some lines may correspond to early precursor cells which have a selective advantage for immortalization or transformation, due to their higher division potential at the moment of oncogene transfer. As few markers specific for early neural precursors are available, the characterization of such lines often remains uncertain. For other lines which do not display clear-cut differentiated properties, it is often assumed that *in vitro* culture conditions lead to de-differentiation. It has been reported that a few passages of some cell lines can result in the loss of differentiation markers (15). The expression of a high level of introduced oncogenes may also inhibit terminal differentiation (43). Using cell lines transformed with a temperature-sensitive oncogene, a temperature shift leading to the inactivation of the oncogene product, has been found to induce differentiation (16). It has also been observed that several cell lines must be growth-arrested at confluence, in order to express terminal differentiation markers (13, 20).

7.2 Induction of differentiation

As might be expected, those culture conditions or agents known to promote the differentiation of various neural cells in primary cultures are also effective in the induction of differentiation of cell lines. The use of chemically defined medium vs. serum-supplemented medium has shown or confirmed the importance of serum-components, for stimulating or inhibiting, or determining the differentiation pathway of cell lines (refs 16 and 17, and Chapter 3). The presence of growth factors or hormones also influences differentiation. For example, bFGF has been shown to accelerate or to promote the differentiation of different lines towards either an astroglial pathway (1), a neuronal pathway (18), or both (15). Moreover, in the case of a sympatho-adrenal precursor cell line, it has been reported that the addition of FGF stimulates the expression of NGF-receptors and that the presence of both FGF and NGF increases the survival and the neuronal differentiation of these cells (18). The addition of dibutyryl-cyclic AMP (db-cAMP) has been shown to stimulate the differentiation of several astroglial cell lines (1, 5).

7.3 Main cell-type specific differentiation markers

The most widely examined markers, for the identification of cell types, are generally those appearing when terminal differentiation has occurred. Astrocytes and neurons can be characterized by the expression of glial fibrillary acidic protein (GFAP) and of neurofilaments (NF), respectively. Monoclonal antibodies against GFAP and the different NF forms are now commercialized (Sigma, Boehringer, for example). Oligodendrocytes are often identified by the immunodetection of the galactocerebroside (Gal-C) with monoclonal antibodies (49) or of proteolipid proteins (PLP) and myelin basic proteins (MBP) with polyclonal antibodies. Another monoclonal antibody, named A2B5, that interacts with the Gq gangliosides was produced against neuronal precursor cells (50). This antibody recognizes various types of mouse neural precursors and neurons (15–17), whereas in the case of rat cells, it interacts preferentially with bipotential oligodendrocyte type 2 astrocyte progenitors (51). Since immunocytochemical procedures depend on whether a marker is present at the cell surface or internally, the fluorescence procedures in *Protocols 13* and *14* are presented for internal, or cell surface labelling.

Protocol 13. Immunofluorescence detection of internal antigens

1. Grow the cells on glass coverslips, coated as described in *Protocols 1–3*.
2. Rinse the coverslips with DMEM or with a balanced salt solution and fix the cells with 4% paraformaldehyde in PBS, for 10 min at RT. Wash the cells briefly with PBS.
3. Permeabilize the cells with acetone at $-20\,°C$, for 10 min and then rinse the cells with PBS. Alternatively, permeabilize the cells with 0.1% Triton X-100 in PBS, at RT.
4. Pre-incubate the cells with non-immune serum (5% in PBS) from the same animal species as that in which the secondary fluorescent antibody (see step 7) has been raised. Wash with PBS supplemented with 5% FCS.
5. Incubate the cells with the antibody against the examined differentiation marker (for instance, GFAP or NF or MBP) at the appropriate dilution in PBS–5% FCS, for 1 h at RT in a humidified box.
6. Wash the cells extensively for 30 min (four changes) with PBS containing 5% FCS.
7. Incubate the cells with the fluorescent secondary antibody at the appropriate dilution in the same buffered solution as above, for 30 min at RT, in the dark.
8. Wash the coverslips extensively as indicated in step 6. Rinse once with water. Wipe the back of the coverslips and mount them on microscope slides in a solution of gelvatol containing 100 mg/ml Dabco (Janssen Chemica) to prevent fluorescence fading (52). Finally, observe the cells with an epifluorescence-microscope.

Protocol 14. Immunodetection of surface markers

1. Prepare cell cultures as indicated in *Protocol 13*, step 1.
2. Rinse the coverslips with DMEM and pre-fix them[a] with 2% (w/v) paraformaldehyde in PBS for 10 min at RT. Wash the cells briefly with PBS.
3. Treat the cells as described in *Protocol 13*, step 4.
4. Incubate the cells with the antibody against the appropriate marker (for instance, Galactocerebroside, Gq ganglioside) appropriately diluted in PBS–5% FCS, for 30 min at RT[a] in a humidified box.
5. Wash the cells as in *Protocol 13*, step 6.
6. For incubation with the secondary conjugated antibody, follow the procedure of *Protocol 13*, step 7. Then, wash the cells extensively, rinsing finally with PBS.

Generation of neural cell lines by transfer of viral oncogenes

Protocol 14. *Continued*

7. Fix in 95% ethanol–5% acetic acid (v/v), for 10 min at −20°C and wash extensively with PBS before mounting the coverslips.

ᵃ Surface antigens can be detected directly on living cells. For this purpose, pre-fixation is omitted and the incubations and washes are carried out at 4°C, in the presence of 10 mM sodium azide.

These procedures can be adapted for double labelling experiments. For example, proceed as follows for the simultaneous detection of Gq ganglioside and GFAP:

(a) Incubate the cells with monoclonal A2B5 antibody as indicated in *Protocol 14* (steps 1–5). Supernatant from hybridoma clone 105 (American Type Culture Collection) can be used as a source of A2B5 antibody.

(b) Fix with 95% ethanol–5% acetic acid (v/v), for 5 min at −20°C. Wash the cells extensively with PBS.

(c) Incubate the cells with rabbit antibodies against GFAP (Sigma, Boehringer) in the conditions described in *Protocol 13*, steps 4–5.

(d) Stain with fluorescein-conjugated goat antibodies against mouse IgM and with rhodamine (or Texas red)-conjugated donkey antibodies against rabbit immunoglobulins, following a procedure similar to that detailed in *Protocol 13*, steps 7–8.

7.4 Other markers and functional properties

Even though the above-mentioned markers are widely used for defining cell types in primary neuronal cultures or immortalized cell lines, other markers may be of interest for characterizing cells. For example, vimentin is an internal marker of some neuronal precursors, although it is not specific for neural cells; the immunodetection of synthesized or secreted laminin can be of interest for determining the astroglial nature of brain-derived cells; the study of the expression of the different forms of N-CAM may also be helpful in the characterization of different types of neural cells (53); see Chapter 9. Ultrastructural studies can provide very useful information. In addition, the measurement of the activity of some enzymes, such as glutamine synthetase, $2',3'$-cyclic nucleotide $3'$-phosphodiesterase, glycerol phosphate dehydrogenase, are informative for determining the astrocytic or oligodendrocytic nature of the cells (32, 33). Functional analysis of cell lines is dealt with in more detail in Chapters 4–9.

8. Future prospects for generating cell lines from transgenic mice

Although the generation of neural cell lines from transgenic mice carrying an oncogene is very recent, and has led to few lines so far (20–22), this approach is likely to become increasingly used. Indeed, tumorigenesis has been targeted towards precise neuronal cells in transgenic mice carrying the SV40 large T gene placed under the control of promoter regions responsible for tissue-specific gene expression. Thus, neuronal cell lines with highly differentiated phenotypes have been derived from tumours induced by SV40 large T gene expression: this expression has been targeted by the promoters of the gonadotropin-releasing hormone gene (22) and of phenylethanolamine *N*-methyl transferase (21). Glial cell lines, able to differentiate *in vitro* have been obtained from non-tumoral brain regions of transgenic mice carrying the polyoma large T gene downstream of its own control region (20). These phenotypically untransformed cells retain contact inhibition of growth. Temperature sensitive oncogenes provide another approach to regulating cell proliferation and differentiation (see Section 4.1.2). Fibroblast and thymic epithelial cell lines have been derived from transgenic mice carrying the SV40 tsA mutant large T downstream from the histocompatibility H-2K promoter, which is partly inducible by interferon-gamma (54). The association of this mutant SV40 gene with neural specific promoters will allow this system to be extended towards different types of neural cells.

Although an introduced oncogene is present throughout the development of the transgenic mouse, and is ready to be expressed at a developmental stage compatible with the promoter associated with the oncogene, the generation of highly differentiated cell lines may be problematic owing to possible post-mitotic expression of genes encoding differentiation markers. For example, it has been reported that the SV40 large T gene controlled by the promoter region of growth hormone releasing factor, led to an unexpected thymic hyperplasia, but did not induce transformation of hypothalamic cells from transgenic mice (55).

Taking into account the small amount of reported data, the following general approach for deriving neural cell lines from transgenic mice suggests itself. First, the regulatory elements of the gene encoding a differentiation marker specific for the cell subpopulation to be immortalized, should be identified and cloned. The specificity of such regulatory sequences could be analysed firstly by transient expression assays with constructs carrying a reporter gene associated with these sequences (see Section 2.1 and Chapter 8). However, such *in vitro* assays are impeded by the limited repertoire of neural cell lines available so far, and by the fact that each line represent a relatively narrow window of development. Thus, the most efficient and reliable method for identifying the regulatory *cis* elements responsible for tissue-specific gene expression, is the introduction of the constructs into the mouse germ-line followed by the study of reporter gene expression in transgenic mice. The methodology of the production of transgenic mice is exhaustively documented (56).

Transgenic mice can then be produced with constructs carrying the same regulatory elements associated with an oncogene instead of a reporter gene.

Depending on the oncogene, the targeted cells may acquire transformed properties and may grow as tumours in the animal, or they may be untransformed, growth-regulated, and located as expected from preliminary experiments with the reporter gene. Whatever the case, the targeted cells should be dissociated from the expected tissue, and the cell lines should be selected on the basis of culture conditions appropriate to each expected cell type (see *Protocols 1–3*). An additional approach to improving the immortalization and the generation of cell lines would be to target both the integration and the expression of an oncogene, by homologous recombination (57, 58) to a region of the genome that can be disrupted without affecting viability (see Chapter 4).

References

1. Evrard, C., Galiana, E., and Rouget, P. (1986). *EMBO J.*, **5**, 3157.
2. Schubert, D., Heineman, S., Carlisle, W., Tarikas, H., Kimes, B., Patrick, J., et al. (1974). *Nature*, **249**, 224.
3. Greene, L. A. and Tischler, A. S. (1976). *Proc. Natl. Acad. Sci. USA*, **73**, 2424.
4. de Vitry, F. (1977). *Nature*, **267**, 48.
5. Mallat, M., Mouro-Neto, V., Gros, F., Glowinski, J., and Prochiantz, A. (1986). *Devel. Brain Res.*, **26**, 23.
6. Kellerman, O. and Kelly, F. (1986). *Differentiation*, **32**, 74.
7. Alliot, F. and Pessac, B. (1984). *Brain Res.*, **306**, 283.
8. Rassoulzadegan, M., Naghashfar, Z., Cowie, A., Carr, A., Grisoni, M., Kamen, R., and Cuzin, F. (1983). *Proc. Natl. Acad. Sci. USA*, **80**, 4354.
9. Ruley, H. E. (1983). *Nature*, **304**, 602.
10. Land, H., Parada, L. F., and Weinberg, R. A. (1983). *Nature*, **304**, 596.
11. Colby, W. W. and Shenk, T. (1982). *Proc. Natl. Acad. Sci. USA*, **79**, 5189.
12. Jat, P. S., Cepko, C. L., Mulligan, R. C., and Sharp, P. A. (1986). *Mol. Cell. Biol.*, **6**, 1204.
13. Evrard, C., Galiana, E., and Rouget, P. (1988). *J. Neurosci. Res.*, **21**, 80.
14. Cepko, C. L., Roberts, B. E., and Mulligan, R. C. (1984). *Cell*, **37**, 1053.
15. Bartlett, P. F., Reid, H. H., Bailey, K. A., and Bernard, O. (1988). *Proc. Natl. Acad. Sci. USA*, **85**, 3255.
16. Frederiksen, K., Jat, P. S., Valtz, N., Levy, D., and McKay, R. (1988). *Neuron*, **1**, 439.
17. Evrard, C., Borde, I., Marin, P., Galiana, E., Premont, J., Gros, F., and Rouget, P. (1990). *Proc. Natl. Acad. Sci. USA*, **87**, 3062.
18. Birren, S. J. and Anderson, D. J. (1990), *Neuron*, **4**, 189.
19. Hammond, D. N., Wainer, B. H., Tonsgard, J. H., and Heller, A. (1986). *Science*, **234**, 1237.
20. Galiana, E., Borde, I., Marin, P., Rassoulzadegan, M., Cuzin, F., Gros, F., Rouget, P., and Evrard, C. (1990). *J. Neurosci. Res.*, **26**, 269.
21. Hammang, J. P., Baetge, E. E., Behringer, R. R., Brinster, R. L., Palmiter, R. D., and Messing, A. (1990). *Neuron*, **4**, 775.
22. Mellon, P. L., Windle, J. J., Goldsmith, P. C., Padula, C. A., Robert, J. L., and Weiner, R. I. (1990). *Neuron*, **5**, 1.

23. Gorman, C., Moffat, L., and Howard, B. (1982). *Mol. Cell. Biol.*, **2**, 1044.
24. Sanes, J. R., Rubenstein, J. L. R., and Nicolas, J-F. (1986). *EMBO J.*, **5**, 3133.
25. Bottenstein, J. E. and Sato, G. H. (1979). *Proc. Natl. Acad. Sci. USA*, **76**, 514.
26. Colbere-Garapin, F., Horodniceanu, F., Kourilsky, P., and Garapin, A. C. (1981). *J. Mol. Biol.*, **150**, 1.
27. Blochlinger, K. and Diggelmann, H. (1984). *Mol. Cell. Biol.*, **4**, 2929.
28. Simonsenn, C. C. and Levinson, A. D. (1983). *Proc. Natl. Acad. Sci. USA*, **80**, 2495.
29. Nicolas, J-F. and Berg, P. (1983). In *Conference on cell proliferation* (ed. L. M. Silver, G. R. Martin, and S. Strickland), Vol. 10, pp. 469–85. Cold Spring Harbor Laboratory Press, Cold Spring Harbor, NY.
30. Cuzin, F. (1984). *Biochem. Biophys. Acta*, **781**, 193.
31. Evrard, C., Ragot, T., Richard-Foy, H., and Rouget, P. (1988). *Genome*, **30** (Suppl. 1), 63.
32. Hallermayer, K., Harmening, C., and Hamprecht, B. (1981). *J. Neurochem.*, **37**, 43.
33. Saneto, R. P. and de Vellis, J. (1985). *Proc. Natl. Acad. Sci. USA*, **82**, 3509.
34. Graham, F. L. and van der Eb, A. J. (1973). *Virology*, **52**, 466.
35. Chen, C. and Okayama, H. (1987). *Mol. Cell. Biol.*, **7**, 2745.
36. Neumann, E., Schaefer-Ridder, M., Wang, Y., and Hofschneider, P. H. (1982). *EMBO J.*, **1**, 841.
37. Blangero, C. and Tessie, J. (1985). *J. Membr. Biol.*, **86**, 247.
38. Chu, G., Hayakawa, H., and Berg, P. (1987). *Nucleic Acids Res.*, **15**, 1311.
39. Behr, J. P., Demeneix, B., Loeffler, J. P., and Mutul, J. P. (1989). *Proc. Natl. Acad. Sci. USA*, **86**, 6982.
40. Felgner, P. L. and Ringold, G. M. (1989). *Nature*, **337**, 387.
41. Weiss, R., Teich, N., Varmus, H., and Coffin, J. (ed.) (1982). *Tumor viruses*. Cold Spring Harbor Laboratory Press, Cold Spring Harbor, NY.
42. Keller, G., Paige, C., Gilboa, E., and Wagner, E. F. (1985). **318**, 149.
43. Trotter, J., Boulter, C. A., Sontheimer, H., Schachner, M., and Wagner, E. F. (1989). *Oncogene*, **4**, 457.
44. Murphy, M., Bernard, O., Reid, K., and Bartlett, P. F. (1991). *J. Neurobiol.*, **22**, 522.
45. Mann, R., Mulligan, R. C., and Baltimore, D. (1983). *Cell*, **33**, 153.
46. Miller, A. D. and Buttimore, C. (1986). *Mol. Cell. Biol.*, **6**, 2895.
47. Danos, O. and Mulligan, R. C. (1988). *Proc. Natl. Acad. Sci. USA*, **85**, 6460.
48. Maniatis, T., Fritsch, E. F., and Sambrook, J. (ed.) (1989). *Molecular cloning. A laboratory manual* (2nd edn). Cold Spring Harbor Laboratory Press, Cold Spring Harbor, NY.
49. Ranscht, B., Clapshaw, P. A., Price, J., Noble, M., and Seifert, W. (1982). *Proc. Natl. Acad. Sci. USA*, **79**, 2709.
50. Eisenbarth, G. S., Walsh, F. S., and Nirenberg, M. (1979). *Proc. Natl. Acad. Sci. USA*, **76**, 4913.
51. Raff, M. C., Miller, R. H., and Noble, M. (1983). *Nature*, **303**, 390.
52. Langanger, G., de Mey, J., and Adam, H. (1983). *Mikroskopie*, **40**, 237.
53. Barbas, J. A., Chaix, J. C., Steinmetz, M., and Goridis, C. (1988). *EMBO J.*, **7**, 625.

54. Jat, P. S., Noble, M. D., Ataliotis, P., Tanaka, Y., Yannoutsos, N., Larsen, L., and Kiossis, D. (1991). *Proc. Natl. Acad. Sci. USA*, **88,** 5096.
55. Botteri, F. M., van der Putten, H., Wong, D. F., Sauvage, C. A., and Evans, R. M. (1987). *Mol. Cell. Biol.*, **7,** 3178.
56. Hogan, B., Constantini, F., and Lacy, E. (1986). *Manipulating the mouse embryo. A laboratory manual.* Cold Spring Harbor Laboratory Press, Cold Spring Harbor, N.Y.
57. Mansour, S. L., Thomas, K. R., and Capecchi, M. R. (1988). *Nature*, **336,** 348.
58. Le Mouellic, H., Lallemand, Y., and Brulet, P. (1990). *Proc. Natl. Acad. Sci. USA*, **87,** 4712.

3

Serum-free media for neuronal cell culture

MICHAEL BUTLER

1. Introduction

The first reported work on the growth of cells *in vitro* was in 1907 by Harrison (1) who was able to suspend an explant of the spinal cord of a frog embryo from a microscope slide. This 'hanging drop technique' enabled the growth of the neuronal cells to take place within a fibrin matrix resulting from the application of a drop of lymph fluid. The observed cell growth was supported by the nutrients provided by the supernatant remaining from the clotted lymph.

Further developments of culture techniques involved the utilization of a variety of biological fluids. Cockerel plasma and chick embryo extracts were widely used because of their good potential to support cell growth. However, inconsistencies in the composition of these biological media caused wide differences in observed cell growth.

In the 1950s, Eagle and others studied the nutritional requirements of mammalian cells in culture (2). The value of developing chemically defined media lay in the ability to grow cells *in vitro* under standardized conditions. Eagle's minimal essential medium (EMEM) was adopted as a standard basal medium which would support the growth of a range of cell lines. Similar culture media such as DMEM, GMEM, Ham's F12, and RPMI 1640 were subsequently developed and have now become widely used formulations for *in vitro* cell culture. However, in most cases cellular growth requirements are not met until a 5 to 10% supplement of blood serum is added to the culture medium.

Blood serum, which is the supernatant of clotted blood contains many ill-defined growth factors essential for cell growth. Fetal calf serum has been found to be particularly efficacious and has been valued for its excellent growth promoting properties. However, its use is associated with several problems:

(a) Batch to batch variation in composition causes inconsistency in growth-promoting properties.

(b) Viral or mycoplasma contamination have been detected in some batches of commercial serum.
(c) The high protein content of serum-supplemented medium can cause difficulties in product extraction.
(d) The presence of serum can mask the effect of hormones or growth factors when studying their interaction with cells.
(e) The availability of fetal calf serum is variable and there have been periods of world shortages.
(f) Fetal calf serum is expensive and can account for up to 85% of the overall cost of medium when calculated for large-scale cultures.

Because of these difficulties various attempts have been made to develop serum-free media formulations which can support cell growth. Early progress was made with tumorigenic cell lines which are less fastidious in their growth requirements. In 1973 for the first time, primary neural tissue was cultured in serum-free medium with the observation of the differentiation of chick neurons in media supplemented with nerve growth factor (3).

More recently, successful serum-free formulations have been developed to support the culture of a range of primary and normal differentiated cells. Some of these formulations have been found particularly useful in the selection of single cell types from the mixed populations obtained from primary sources and for the study of the effect of factors on neural cell differentiation.

2. The development of serum-free media

The function of a serum supplement in culture can be listed as follows:

(a) To promote cell attachment to the available substratum. This is particularly important for anchorage-dependent cells at the early stage of culture.
(b) To provide nutrients, particularly those required in trace concentrations.
(c) To provide growth stimulation. This function may be associated with growth factors, hormones, or enzymes present in the serum. Serum may also contain regulators of membrane permeability or passive carriers of micro-nutrients, either of which may result in growth promotion.
(d) To protect cells from damage which may be mechanical (such as shear) or biological (for example, enzymes).

Various approaches have been attempted to understand these requirements for serum in order to develop a systematic approach of substituting serum by chemically defined components.

2.1 The analytical approach

Attempts have been made to analyse the composition of serum and its usage during culture with a view to developing a synthetic media from the selected

Table 1. Some major plasma protein types

Protein	Concentration in human plasma (g/litre)
Albumin	35–55
Immunoglobulin G	8–18
Immunoglobulin A	0.8–4.5
Fibrinogen	3
Transferrin	3
α_1-antiprotease	3
α_2-macroglobulin	2.5
Immunoglobulin M	0.6–2.5
Haptoglobulin	1.7–2.3
Fibronectin	0.3

From ref. 4

and identified components. The major groups of plasma proteins are listed in *Table 1* and *Figure 1* shows the results of an electrophoretic analysis of the serum supplement of media before and after cell growth (5). The lower content of α-2 and β fractions indicates that these may be absorbed into the cells. Theoretically, it should be possible to develop a defined medium after identification of the active components in these fractions.

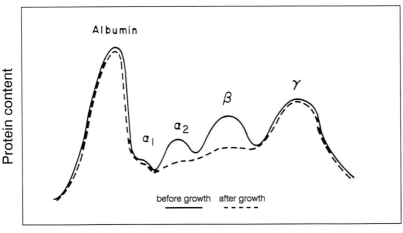

Figure 1. Electrophoresis of serum proteins extracted from culture medium of rat tumour cells. (From ref. 5.)

Table 2. Typical range of mitogenic stimulation of a single component in serum

Serum supplement in medium = 10–20%
Medium protein content = 6–12 mg/ml
Activity range of known mitogens = 10^{-10}–10^{-8} M
Assume a molecular weight = 100 kd
Effective concentration of mitogen = 10^{-5}–10^{-3} g/litre
Thus % of mitogen in serum protein = 2×10^{-2}–5×10^{-6}%

Although some attention has been given to such techniques in attempting to analyse and characterize the essential individual serum proteins, there are fundamental difficulties in this approach. Some proteins may be required during growth but are not absorbed into the cells. Furthermore, each active component may exert its effect at extremely low concentrations. *Table 2* indicates a typical range of mitogenic activity and shows that an individual mitogen may account for as little as 5×10^{-6}% of the serum protein. Observation and identification of the effect of an individual mitogen or growth stimulator at such low concentrations would be extremely difficult. Such difficulty may be compounded by its synergistic activity in relation to other components present in the serum.

2.2 Ham's approach

Richard Ham has developed a systematic and meticulous approach to the development of serum-free media by the gradual reduction in the concentration of undefined supplements in the culture (6).

This approach is based upon two propositions:

(a) A complex interaction exists between the components of culture media to meet cell growth requirements. Thus, a change in the concentration of one component might have an effect upon the optimal concentration of another.

(b) Under suboptimal growth conditions, the rate of cell growth is dependent upon one limiting factor ('the first limiting factor').

Applying these propositions, Ham developed a systematic method for the development of defined media (*Protocol 1*). This is a time-consuming operation which is based initially upon optimization of the concentration of nutrients in the basal medium. For most nutrients there is a wide concentration range over which the nutrient is not growth limiting. This range can be established by adding the nutrient to media at concentrations of ×0.1, ×1, ×10, ×100, etc., of the original concentration. By comparing the growth response at each concentration, the mid-point of the log scale reflecting the optimum concen-

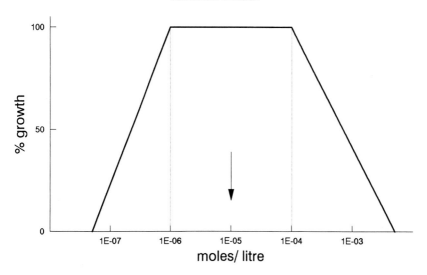

Figure 2. Idealized growth response curve for a single nutrient.

tration range can be determined (*Figure 2*). If the nutrient concentration is set near the centre of the broad optimum plateau, then it is less likely to become growth limiting by adjustments of other media components. In the growth response curve shown in *Figure 2* the centre of the optimum concentration is 10^{-5} M.

This procedure of optimization of the basal media may reduce the dependence on a serum supplement to a low level (about 2%). Cells can adapt to reduced serum by the stimulation of anabolic processes. It may be necessary to allow the cells to grow for several generations at each step of serum reduction to allow the required adaptation to become established.

The ability of cells to grow in a serum content of less than 2% will vary and may depend upon the availability of growth factors and hormones to replace those normally supplied by the serum. A dialysis step is useful in order to distinguish between the requirements for low- and high-molecular-weight factors. The addition of selected components to cells growing at 50% maximal rates allows an easy method of detecting growth enhancement.

Protocol 1. Ham's approach to the design of serum-free medium

1. Use any means necessary to obtain growth of the cells of interest. This may include serum supplementation, the use of feeder layers or of conditioned media.
2. Select the combination of readily available media to obtain the maximum cell growth.

Protocol 1. *Continued*

3. Replace all undefined supplements with dialysed supplements. Add the undefined low-molecular-weight supplements which were removed by dialysis.
4. Identify all the low-molecular-weight supplements required for growth by a combination of analysis and trial additions.
5. Reduce the addition of dialysed supplements to a level that supports less than maximal growth (for example, 50% of maximum). Thus, the undefined supplements are now rate-limiting to cell growth.
6. Sequentially adjust the concentrations of all defined nutrient components of the basal media to experimentally determined optima.
7. Sequentially test defined low-molecular-weight substances for growth enhancement.
8. At each stage of growth improvement, reduce the amount of undefined supplement so that it becomes rate-limiting.
9. Repeat the procedures in steps 7–8 above until no further reduction in the amount of dialysed supplement is possible without affecting cell growth.

2.3 Sato's approach

Gordon Sato and his colleagues have devoted their efforts to understanding the interaction between cells and growth factors or hormones (7).

They argue that the survival of cells in culture is dependent upon the ability of media components to perform the functions previously provided by the normal environment of the cell *in vivo*. Thus, the design of defined media should be based upon an attempt to re-construct the extracellular environment involved in supporting cell growth *in vivo*. This environment involves a combination of hormones, growth factors, binding proteins, and attachment factors. It is argued that although serum supplements may support some cell growth, serum may be an unsuitable substitute for the *in vivo* environment in many cases. Serum contains substances which may never come in contact with some cells *in vivo*, and the toxicity of high levels of serum has been reported in many instances. This includes reports of the presence in serum of inhibitors of neural differentiation (8).

Thus, the Sato approach involves the formulation of chemically defined media for optimization of cell growth from a list of isolated and purified growth-promoting substances. These formulations are not perceived as substitutes for serum supplementation but rather an attempt to simulate the *in vivo* environment.

Advantages of media supplemented with defined growth factors and hormones include:

(a) In many cases cells show enhanced growth characteristics in these defined media compared to serum-containing media. Some cells notably differentiated types, cannot be maintained at all in serum-supplemented media and their maintenance *in vitro* is dependent upon selected hormones and growth factors.

(b) It is possible to select specific cell types from a mixed population of cells as may be obtained from a primary source. Such cell-specific selection requires careful manipulation of the media composition.

(c) Studies of the interaction of hormones or drugs on cultured cells are made possible by such media. The uncertain composition of serum and the binding effects of the associated proteins would limit such studies in serum-supplemented media.

3. Defined media supplements

The chemically defined supplements which are added to the basal media to promote cell growth or maintenance are required in relatively low concentrations to stimulate cellular activity. They are not nutrients in the sense that they are not utilized as substrates for metabolic pathways. Normally, a defined composition of these components would be added to basal media as a predetermined cocktail which would obviate the need for serum supplementation.

These supplements which can be obtained commercially (for example, Sigma) can be divided into several groups as follows.

3.1 Hormones

These are generally low-molecular-weight compounds which have well-characterized regulatory functions *in vivo*. *In vitro*, hormones show specific effects depending upon the nature of the cultured cells. The addition of some hormones to culture media has been shown to be growth stimulatory for a range of cell lines. These include:

3.1.1 Insulin

Insulin appears to be a universal requirement for the *in vitro* growth of cells and is included in most serum-free media formulations. This is a relatively small polypeptide (M_r 5.7 kDa) that is known to have multiple effects on cell metabolism. In particular, cell membrane transport, glucose metabolism, fatty acid biosynthesis, and nucleic acid biosynthesis have all been shown to be stimulated in the presence of insulin (9).

The stimulation of DNA synthesis by insulin has been shown to be particularly important in the stimulation of cell growth *in vitro*. Experiments have shown the ability of insulin to stimulate the uptake of tritiated thymidine into nuclei and the transformation of cells from gap G_1 phase into the synthetic S-phase during the mitotic cycle.

It is notable that the insulin concentration often found to be optimal in cell culture (about 5 μg/ml) is approximately a thousand times higher than that found in serum. The need for this higher concentration may be related to the rate of inactivation of free insulin in growth medium. It has been shown that 90% of the insulin activity can be destroyed in some media in 1 h at 37°C (10). This is likely to be due to the reduction of the disulfide bridges in insulin by agents such as cysteine in the medium. Replacement of cysteine by cystine can reduce the inactivation process. Similarly, the presence of some serum can stabilize insulin and allow lower concentrations to stimulate cell growth.

Commercially available insulin includes the extracted hormone from bovine or porcine pancreas and the human recombinant form. Insulin is most stable in acidic solutions which should be used to reconstitute insulin solutions from a lyophilized powder. The biological activity of insulin is dependent on zinc which should be present in the supplement (at about 15 μM) prior to addition to the culture medium. Some commercial sources of insulin are supplied in a formulation containing zinc.

3.1.2 Transferrin

This protein (M_r 86 kDa) is a constituent of the β-globulin fraction of blood serum and has been shown to be important for iron transport under normal physiological conditions. The inclusion of transferrin in most serum-free formulations appears to be related to its ability to transport iron into the cell, although it may also act to detoxify trace metals present in the medium.

The close relationship between transferrin and iron in growth stimulation was demonstrated in an experiment (11) in which the observed effect of transferrin on cells was shown to be dependent upon the presence or absence of $FeSO_4$ in the medium (*Figure 3*). In the absence of iron there was a 15-fold increase in cell growth at a transferrin concentration of 1 μg/ml. However, in the presence of iron the initial growth rate was higher but the same transferrin concentration increased cell growth by only ×2.

Iron-saturated transferrin is available commercially and may be used in preference to the iron-free form which would require the presence of iron in the medium for maximum stimulation.

3.1.3 Steroid hormones

A range of glucocorticoid steroids show growth stimulation within a concentration range around 10^{-8} M which is the concentration found in the bloodstream under physiological conditions. Testosterone, progesterone, and dexamethasone have been shown to be individually stimulatory in serum-free medium but their effects are not additive in the presence of optimal concentrations of other hormones. It is suggested that the activity of these steroids is elicited by a common metabolic product.

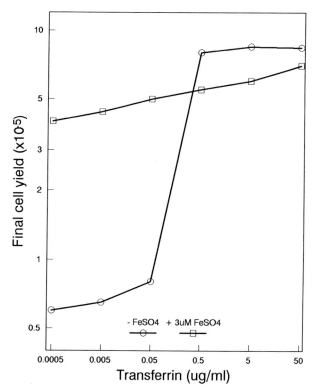

Figure 3. The effect of transferrin on the growth of mouse melanoma cells in Ham's F-12 medium. (From ref. 11.)

The synthetic glucocorticoid, dexamethasone, enhances the mitogenic action of insulin and epidermal growth factor by increasing the binding capacity of the corresponding membrane receptors. A similar action is displayed by hydrocortisone which has also been shown to promote spreading of fibroblasts on a substratum (12, 13).

3.2 Growth factors

Growth factors are proteins originally derived by extraction from specific mammalian tissue, many originating from endocrine glands. However, some are now available as recombinant proteins. The latter are favoured in media formulations because of their high purity. In some cases the physiological effects of growth factors are not completely characterized, but they do have mitogenic and growth stimulatory effects on cells *in vitro*. Growth factors have one or more of the following effects on cultured cells:

- initiate DNA synthesis
- stimulate one cycle of division in confluent cultures

- stimulate several cycles of division in sparse cultures
- stimulate clonal growth

3.2.1 Epidermal growth factor (EGF)

This is a low-molecular-weight (6.1 kDa) acidic protein originally isolated from mouse submaxillary gland which appears to be a rich source of this factor. It can be extracted from human urine (urogastrone), which has been found to be effective in the growth enhancement of human diploid cells. It is also available as a human recombinant protein expressed in *Escherichia coli*. The recombinant EGF has a lower molecular weight and has two amino acids missing at the C-terminal but has indistinguishable properties from the naturally occurring molecule.

EGF has been found to modulate neuronal differentiation and to be a potent mitogen for primary cultures and particularly fibroblast, epidermal, and glial cells. Astrocytes from primary brain tissue have shown increased proliferation rates in the presence of EGF (14).

Cells that are transformed tend to loose their requirement for EGF. In serum, EGF circulates as a complex bound to several carrier proteins. Its mitogenic properties relate to the large number of receptor sites found on the cell membrane. Steroids have been found to have a synergistic effect on EGF activity. Recommended range of media supplementation: 1–20 ng/ml.

3.2.2 Fibroblast growth factors (FGF)

This group of polypeptides was first isolated from bovine pituitary glands and subsequently from bovine brain tissue (15). Isolates from both sources stimulate the same range of fibroblastic cells, despite little structural resemblance. Brain FGF can be separated into acidic-FGF and basic-FGF which have 55% homology.

Basic-FGF is a low-molecular-weight protein (16–18 kDa). This has been found to be a potent mitogen for a range of cell types (not just fibroblasts). The mitogenic activity acts through the stimulation of calcium influx and the activation of protein kinase. It can induce the synthesis of collagen and may become incorporated into the extracellular protein matrix.

- Recommended range of media supplementation: 0.5–2 ng/ml.

Acidic-FGF is less potent as a mitogen than basic-FGF although its activity may be enhanced by heparin which stabilizes its tertiary structure. This has also been designated 'endothelial cell growth factor' because of its mitogenic effect on endothelial cells.

- Recommended range of media supplementation: 100–300 μg/ml (or 0.5–10 ng/ml with heparin).

3.2.3 Gimmel factor

This factor can be isolated from rat submaxillary gland and has been shown to be mitogenic for glioma cells. Although closely associated with the activity of EGF and FGF it has been shown to be a distinct factor (16).

- Recommended range of media supplementation: 10–15 µg/ml.

3.2.4 Nerve growth factor (NGF)

This was one of the first growth factors to be described (17). NGF is a dimer of two identical polypeptide chains (M_r 26.5 kDa). It is not a potent mitogen but it can stimulate the secretory activity and differentiation of cells of the peripheral nerve system. NGF can be purified from mouse submaxillary glands where it is found at high concentrations. This is available as a high-molecular-weight complex (130 kDa and designated 7S-NGF) or as a purified dipeptide (designated 2.5S-NGF). It circulates in serum bound to a carrier protein which is commercially available as a 140-kDa serum complex, which is stable but has low activity. The purified polypeptide dimer has a higher specific activity but is less stable.

- Recommended range of media supplementation: 5–10 ng/ml.

3.3 Attachment factors

A number of proteins are important to ensure the attachment of anchorage-dependent cells to their substratum. The attachment proteins (also referred to as extracellular matrix, ECM proteins) are present in serum and can also be secreted by cells.

3.3.1 Fibronectin

Fibronectin is a high-molecular-weight multimeric glycoprotein found on the surface membrane of cells. There are two forms of fibronectin that are generally available. Plasma fibronectin (dimer with 200–220 kDa subunits) has been isolated from blood plasma and also been termed cold-insoluble globin (CIg). Cellular fibronectin (matrix of 2200–2500 kDa subunits) has been isolated from the surface of fibroblast cells and has in the past, also been termed large-external-transformation-sensitive (LETS) protein or cell-surface protein (CSP).

In cultures, fibronectin is important in cell-substratum adhesion of anchorage-dependent cells. As cells synthesize fibronectin, the need for media supplementation is dependent on the cell type. Transformed cells lose the ability to produce fibronectin and this may partly explain their loss of anchorage dependence. Many cell types show improved adherence and growth on fibronectin-coated surfaces (18).

- Recommended concentration for coating substratum: 5–20 µg/ml.

3.3.2 Laminin

This is an attachment protein which may be secreted by some cells or may be added to culture dishes to enhance cell attachment (19). Laminin also enhances the growth of human fetal sensory neurons *in vitro* (20).

- Culture vessel coating is recommended at 2–5 µg/cm^2.
- Recommended concentration for coating substratum: 20 µg/ml.

3.3.3 Vitronectin
This is an acidic glycoprotein (M_r 78 kDa) with a high binding affinity for glass. It can be extracted from serum and was originally designated 'serum spreading factor' because of its ability to promote cell adhesion.

3.4 Carrier proteins
These are important to promote the transport of essential nutrients or trace elements. Such components may also be classed as hormones; for example, transferrin. Albumin is often incorporated into serum-free formulations and may serve as a carrier for lipids or trace elements.

3.5 Lipids
Lipids have a variety of functions which include serving as a structural component of membranes and as a stored source of metabolic fuel. Serum is a good source of lipids which are rarely included in basal media. Specific lipids may be included as supplements in serum-free medium and have been shown to improve cell proliferation (21).

Serum-free formulations often include a source of essential fatty acids, phospholipids, lecithin, or cholesterol. Fatty acids may be dependent on the presence of a larger carrier molecule such as albumin for access to cells. Arachidonic acid and docosahexaenoic acid are polyunsaturated fatty acids found in neuronal cell membranes and their addition to serum-free media can allow the normal synthesis of cell membranes (22).

Prostaglandins have also been found to enhance cell growth and may act synergistically with growth factors (for example, EGF).

3.6 Trace elements
Trace quantities of inorganic elements which are not present in basal media and are present in serum are sometimes included in serum-free formulations. Selenium is the most notable example. This is commonly added to serum-free formulations as sodium selenite and is required to activate specific enzymes involved in metabolic detoxification.

The inclusion of inorganic salts can sometimes reduce the requirement for protein factors as in the case of ferrous salts and transferrin (see *Figure 3*). It is possible to create a totally protein- and peptide-free media formulation for some cells by liberal additions of trace elements as reported for hybridoma cell cultures (23).

4. Serum-free formulations for specific neural cell types

There is no serum-free formulation which has been found applicable to all cell lines. Each cell type has particular requirements which have to be determined by systematic techniques as outlined earlier. However, it would seem that insulin, transferrin, and selenium are three components which have universal applications and are included in the majority of serum-free media.

Tables 3 and *4* contain details of various serum-free formulations that have been used for growing or maintaining specific types of neural cells. These formulations vary in their complexity and this is a reflection of the purpose for which each was originally designed. For example, medium 3 (*Table 3*) which contains 13 ingredients was optimized to promote specific cholinergic differentiation of neurons isolated from fetal rat brain. On the other hand, simpler media (such as medium 11) can be used for the proliferation of glial cells.

Media formulations have also been developed for the growth and differentiation of re-aggregated cells in a tissue-type of matrix. An example is the R-16 media which was designed for the culture of neonatal rat neocortex tissue in an organotypic form. The medium formulation, which contains 16 components, needs to be complex to accommodate the requirements of differing cell types contained in the tissue (36).

4.1 Neuronal cell lines

Studies with rat neuroblastoma cells resulted in the development of a serum-free medium (24) (designated 'N2') containing bovine insulin, human transferrin, progesterone, and putrescine (see *Tables 3* and *4*). These promote the growth of a range of neuronal cell lines in a basal medium of Ham's F12 and DMEM (1:1). However, the optimal transferrin concentration may be different in each case, within the range 5–100 µg/ml.

Although the cell growth rate in this media is often slightly lower than that in serum-based media, the final cell density is equivalent. The N2 medium can support the growth of neuroblastoma cells from low densities as well as during serial subculture. Furthermore, cryo-preservation of the cells in N2 medium containing 10% DMSO has been shown possible.

In order to utilize this medium in the absence of serum, culture dishes are coated with poly-D-lysine and fibronectin. *Protocol 2* indicates the steps necessary for this. The culture surface must be pre-coated initially with polylysine, i.e. before addition of media. Fibronectin can be included in the medium which is added to the culture vessel prior to inoculation of cells. The fibronectin rapidly adheres to the culture surface. A synergistic effect has been observed with polylysine and fibronectin treatment, both of which are necessary to ensure cell attachment and growth.

Table 3. Serum-free formulations used for cells derived from neuronal tissue

Component	Media (see Table 4 for details)													
	1	2	3	4	5	6	7	8	9	10	11	12	13	14
Insulin (µg/ml)	5	5	5	–	25	5	2	5	5	50	5	5	25	10
Transferrin (µg/ml)	100	100	1	50	100	100	5	100	10	–	50	0.5	25	100
Progesterone (nM)	20	20×10^3	–	–	20	20	–	–	–	–	–	–	–	–
17-β-Estradiol (nM)	–	–	–	–	–	10^{-3}	–	36.8	–	–	–	–	–	–
Corticosterone (nM)	–	–	–	–	–	100	–	–	–	–	–	–	–	–
Hydrocortisone (nM)	–	20	20	10	–	–	–	–	–	50	–	–	–	–
Triiodothyronine (nM)	–	0.03	30	–	–	10	–	0.03	–	–	–	–	–	–
Thyroglobulin (µg/ml)	–	100	–	–	–	–	–	100	–	–	–	–	–	–
EGF (ng/ml)	–	–	–	–	–	–	–	–	–	10	–	–	–	1.6
FGF (ng/ml)	–	–	–	5	–	–	–	–	–	100	–	75	50	–
NGF (µg/ml)	–	–	5	–	–	–	–	–	–	–	–	–	–	–
Gimmel factor (µg/ml)	–	–	–	–	–	–	10	–	–	–	–	–	–	–
Putrescine (µM)	100	100	–	–	60	100	–	–	–	100	–	–	–	–

Component												
Hyaluronic acid (μg/ml)	–	–	–	–	–	–	–	–	–	–	–	10
Vitamin B$_{12}$ (μg/ml)	–	–	1.4	–	–	–	–	–	–	–	–	–
Biotin (μg/ml)	–	–	1	0.01	–	–	–	–	–	0.01	–	–
α-Tocopherol (μg/ml)	–	–	10	–	–	–	–	–	–	–	–	–
Retinol (μg/ml)	–	–	5	–	–	–	–	–	–	–	–	–
Choline (mM)	–	–	1	–	–	–	–	–	–	–	–	–
Carnitine (uM)	–	–	10	–	–	–	–	–	–	–	–	–
Arachidonic acid (μg/ml)	–	–	–	–	1.0	–	–	–	–	–	–	–
Docosahexaenoic acid (μg/ml)	–	–	–	–	0.5	–	–	–	–	–	–	–
Linoleic acid (μM)	–	–	10	–	–	–	–	–	–	–	17.9	–
Lipoic acid (μM)	–	–	1	–	–	–	–	–	–	–	–	–
Prostaglandin F$_{2α}$ (μg/ml)	–	–	–	–	–	–	–	0.5	–	–	–	–
BSA (mg/ml)	–	–	–	–	0.04	–	1	–	–	–	–	1
Trypsin inhibitor aprotinin (μg/ml)	–	–	–	–	–	–	–	–	–	–	–	1
Sodium selenite (nM)	30	30	–	30	30	30	–	–	–	30	–	30

Table 4. Serum-free media characteristics

Medium (from Table 3)	Reference	Designated name	Basal medium	Type of cells cultured
1	24	N2	DMEM/Ham's 12 (1:1)	Neuroblastoma (rat)
2	25	B17	DMEM/Ham's F12 (1:1)	Hypothalamus (rat)
3	26		DMEM	Fetal brain (rat)
4	27	G2	DMEM	Glial cells (rat)
5	28	N2 (modified)	DMEM/Ham's F12 (1:1)	Neurons (rat)
6	22	SF3	DMEM/Ham's F12 (1:1)	Neurons (mouse hypothalamus)
7	29	C_0	DMEM/Ham's F12 (1:1)	Astrocytes (variants, rat)
8	29	B_9	DMEM/Ham's F12 (1:1)	Astrocytes (selection, rat)
9	30		Waymouth MD 705/1	Astrocytes (rat)
10	31		DMEM/Ham's F12 (1:1)	Astrocytes (rat)
11	32	O1	DMEM	Oligodendrocytes (rat)
12	33		DMEM/Ham's F12 (1:1)	Oligodendrocytes (rat)
13	34	C_6	DMEM/Ham's F12 (3:1)	Glioma cell line (rat)
14	35	A	BMEM–Earle	Astrocytes (mouse)

Protocol 2. Treatment of growth surface for neural cell growth

1. Add a solution of poly-D-lysine (0.05 mg/ml) to a polystyrene T-flask for 5 min at room temperature. An 0.5-ml solution per 8 cm^2 surface area is sufficient.
2. Wash with sterile water at 2 ml per 8 cm^2.
3. Add fibronectin to the basal medium (Ham's F12/DMEM 1:1) at a concentration of 1–10 µg/ml. This is sufficient to coat the surface with fibronectin, providing the medium is added to the T-flask prior to cell inoculation.

4.2 Primary neurons

The N2 medium with a low transferrin concentration and added nerve growth factor can support the survival and expression of many physiological and biochemical properties characteristic of differentiated neurons derived from primary tissue. Specific examples include the selective growth of post-mitotic neurons from chick embryos (37) and the long-term maintenance of embryonic mouse dorsal root ganglion neurons (38).

Using such methods, neurons can be selectively cultured from fetal brain tissue in preference to glial cells which may be present in the initial tissue sample (28). The medium (No. 5 in *Table 3*) is a serum-free formulation modified from N2 and is an alternative for neuron cell selection.

Anti-mitotic agents such as cytosine arabinoside may be used to completely eliminate dividing glial populations, although such agents do exhibit some neuronal toxicity (39).

Medium 3 listed in *Table 3* is a serum-free formulation which was developed to stimulate cholinergic differentiation in neurons isolated from fetal rat brain cells. The medium contains a range of hormones which induce particular biochemical markers. In particular, the presence of triiodothyronine in conjunction with nerve growth factor enhances the irreversible development of choline acetyltransferase (CAT) activity (26). Phenotypic expression of isolated neuronal cells can also be observed in other media listed in *Table 3* (22).

Primary cultures usually require treatment of the growth surface with serum or collagen before cell attachment occurs. Unlike neuroblastoma cell lines, polylysine and fibronectin is not effective for such primary cells.

A novel technique involves three-dimensional rotation cultures (25). With this technique constant rotation of the culture vessel induces the cells to grow as aggregates which have the characteristics of reconstituted tissue. Hypothalamic 'organoids' have been shown to secrete vasopressin into culture medium for several weeks.

4.3 Glial cell lines

A serum-free formulation (designated 'G2') and incorporated into GMEM as the basal medium can support the growth of glial cells with biochemical markers characteristic of mature astrocytes (27).

To allow cell attachment to the substratum, a surface treatment with polylysine and fibronectin is necessary as outlined in *Protocol 2*. These cells are unusual in that they do not have a requirement for insulin which is not included in the G2 medium.

4.4 Primary glial cells

Astrocytes can be selectively grown from brain tissue by the G2 medium (30, 40).

The addition of insulin and epidermal growth factor is required for prolonged culture and survival of these cells (31). EGF may stimulate cellular fibronectin synthesis and removes the need for its exogenous addition.

Carefully designed serum-free formulations can also be used to select variants of astroglial cells (29, 35). Oligodendrocytes can be isolated from primary neonatal brain tissue and be selectively cultured in serum-free formulations (refs 32, 33, and Chapter 2).

5. Undefined low protein serum substitutes

Extracts of serum or plasma may serve to replace the growth-promoting activities of a complete serum supplement. This approach reduces the protein content of the medium and consequently some of the batch to batch variation associated with serum. The exact chemical composition of these extracts is unknown, however.

Cohn fractionation of blood is a method used to separate plasma fractions by sequential steps of ethanol precipitation. This method is used for the large-scale preparation of haemophilic factors from pooled human blood. Cohn fraction 4.1 which is a by-product of this process has been found to possess growth-promoting properties which can substitute for whole serum in growth medium (41).

Low protein serum extracts are available commercially as a range of serum substitutes some of which are listed in *Table 5*.

6. Commercially available serum substitutes

There are a number of commercially available formulations which are designed as complete serum-free media for cell culture or as substitutes for serum in media supplementation. In most cases the formulations are chemically defined but the complete compositions are proprietary. In other cases, undefined extracts of biological fluids are used. Some of these are listed in *Table 5*.

These media formulations have been tested on specific cell lines and in most cases experimental data is available from each company to show the cell growth promoting qualities of the appropriate product compared to a serum-based medium. Although some of the formulations have been developed specifically for hybridoma cultures, they may also have applications for other cell lines of interest.

7. Conclusion

Considerable advances have been made in recent years to define media for specific cells of neural origin. This has allowed the selection of particular cell types from neural tissue. Furthermore, the proliferation of such cells in defined media enables studies of the effects of diffusible trophic or regulatory factors. Such culture systems have become valuable tools which may lead to a

Table 5. Commercially available serum-free media and serum substitutes

Company	Product name	Defined or undefined
Serum-free media (complete)		
Ventrex	HL-1	D
Flow	BioRich	D
Gibco	Hybridoma-SFM	D
Gibco	PFHM II	D
Serum substitutes		
Boehringer Mannheim	Nutridoma	D
Collaborative Res.	Nu-Serum	U
″	ITS and ITS+	D
″	MITO	D
IBF Biotechnics	Ultroser G/HY	D
Dextran Products	CLEX	U
Sigma	CPSR	U
″	Seru-Max	U
″	LPSR	D
Gibco	OptiMab	D
Miles Inc.	Ex-Cyte	U

Abbreviations: CPSR = Controlled Process Serum Replacements: LPSR = Low Protein Serum Replacements; ITS = Insulin/Transferrin/Selenite; PFHM = Protein-free Hybridoma Medium

greater understanding of some of the complex processes associated with neural cell–cell interaction and cell differentiation.

References

1. Harrison, R. (1907). *Anat. Rec.,* **1,** 116.
2. Eagle, H. (1959). *Science,* **130,** 432.
3. Luduena, M. A. (1973). *Devel. Biol.,* **33,** 470.
4. MacLeod, A. J. (1987). *Adv. Biochem. Eng.,* **37,** 41.
5. Kent, H. N. and Gey, G. O. (1960). *Science,* **131,** 666.
6. Ham, R. G. (1983). In *Hormonally defined media* (ed. G. Fischer and R. J. Wieser), pp. 16–30. Springer-Verlag, Berlin.
7. Barnes, D. and Sato, G. (1980). *Cell,* **22,** 649.
8. Davis, G. E., Skaper, S. F., Manthorpe, M., Moonen, G., and Varon, S. (1984). *J. Neurosci. Res.,* **12,** 29.
9. Komolov, I. S. (1978). *Endocrinol. Exp.,* **12,** 43.

10. Hayashi, I., Larner, J., and Sato, G. (1978). *In Vitro,* **14,** 24.
11. Mather, J. P. and Sato, G. H. (1979). *Exp. Cell Res.,* **120,** 191.
12. Baker, J. B., Barsh, G. S., Carney, D. H., and Cunningham, D. D. (1978). *Proc. Natl. Acad. Sci. USA,* **75,** 1882.
13. Hoshi, H., Kan, M., Yamane, I., and Minamoto, Y. (1982). *Biomed. Res.,* **3,** 546.
14. Leutz, A. and Schachner, M. (1981). *Cell Tissue Res.,* **220,** 393.
15. Gospodarowicz, D., Bialecki, H., and Greenburg, G. (1978). *J. Biol. Chem.,* **253,** 3736.
16. McClure, D. B., Ohasa, S., and Sato, G. H. (1981). *J. Cell Physiol.,* **107,** 195.
17. Bradshaw, R. A. (1978). *Ann. Rev. Biochem.,* **47,** 191.
18. Orly, J. and Sato, G. (1979). *Cell,* **17,** 295.
19. von der Mark, K. and Kuhl, U. (1985). *Biochim. Biophys. Acta,* **823,** 147.
20. Baron-Van Evercooren, A., Kelinman, H., Ohno, S., Marangos, P., Schwartz, J., and Dubois-Dalcq, M. (1982). *J. Neurosci. Res.,* **8,** 179.
21. Glassy, M. C., Tharakan, J. P., and Chau, P. C. (1988). *Biotechnol. Bioeng.,* **32,** 1015.
22. Tixier-Vidal, A. and Faivre-Bauman, A. (1990). In *Cell culture: methods in neurosciences* (ed. P. M. Conn), Vol. 2, pp. 355–71. Academic Press, San Diego.
23. Cleveland, W. L., Wood, I., and Erlanger, B. F. (1983). *J. Immunol. Methods,* **56,** 221.
24. Bottenstein, J., Hayashi, I., Hutchings, S., Masui, H., Mather, J., McClure, D. B., et al. (1979). *Methods in enzymology,* Vol. 53, p. 94. Academic Press, Orlando, Florida.
25. Lang, E., Lang, K., Krause, U., Racke, K., Nitzgen, B., and Brunner, G. (1983). In *Hormonally defined media* (ed. G. Fischer and R. J. Weiser), pp. 250–63. Springer-Verlag, Berlin.
26. Honegger, P. and Guntert, B. (1983). In *Hormonally defined media* (ed. G. Fischer and R. J. Wieser), pp. 203–14. Springer-Verlag, Berlin.
27. Michler-Stuke, A. and Bottenstein, J. (1982). *J. Neurosci. Res.,* **7,** 215.
28. Dreyfus, C. and Black, I. B. (1990). In *Cell culture: methods in neurosciences* (ed. P. M. Conn), Vol. 2, pp. 3–16. Academic Press, San Diego.
29. Lang, K. and Brunner, G. (1983). In *Hormonally defined media* (ed. G. Fischer and R. J. Wieser), pp. 222–4. Springer-Verlag, Berlin.
30. Weibel, M., Pettmann, B., Daune, G., Labourdette, G., and Sensenbrenner, M. (1983). In *Hormonally defined media* (ed. G. Fischer and R. J. Wieser), pp. 229–33. Springer-Verlag, Berlin.
31. Morrison, R. S. and Vellis, J. de (1984). In *Methods for serum-free culture of neuronal and lymphoid cells* (ed. D. W. Barnes, D. A. Sirbasku, and G. H. Sato), pp. 15–22. Alan R. Liss, New York.
32. Bottenstein, J. E. and Hunter, S. F. (1990). In *Cell culture: methods in neurosciences* (ed. P. M. Conn), Vol. 2, pp. 56–75. Academic Press, San Diego.
33. Saneto, R. P. and Vellis, J. de (1985). *Proc. Natl. Acad. Sci. USA,* **82,** 3509.
34. Wolfe, R. A., Sato, G., and McClure, D. B. (1980). *J. Cell Biol.,* **87,** 434.
35. Fischer, G. (1983). In *Hormonally defined media* (ed. G. Fischer and R. J. Wieser), pp. 189–202. Springer-Verlag, Berlin.
36. Romijn, H. J., De Jong, B. M., and Ruijter, J. M. (1988). *J. Neurosci. Methods,* **23,** 75.
37. Bottenstein, J., Skaper, S., Varon, S., and Sato, G. (1980). *Exp. Cell Res.,* **125,** 183.

38. Bottenstein, J. E. (1985). In *Cell culture in the neurosciences* (ed. J. E. Bottenstein and G. Sato), pp. 3–43. Plenum Press, New York.
39. Wallace, T. L. and Johnson, E. M. (1989). *J. Neurosci.*, **9**, 115.
40. Michler-Stuke, A., Wolff, J., and Bottenstein, J. (1984). *Int. J. Devel. Neurosci.*, **2**, 575.
41. MacLeod, A. J. (1991). In *Mammalian cell biotechnology: a practical approach* (ed. M. Butler), pp. 28–38. Oxford University Press, Oxford.

4

Embryonal carcinoma cells and embryonic stem cells as models for neuronal development and function

JAMES W. McCARRICK and PETER W. ANDREWS

1. Introduction

For many years teratocarcinomas and their stem cells, embryonal carcinoma (EC) cells, have been well established tools for investigating cell differentiation as it relates to early embryogenesis, especially in the mouse (1, 2). More recently, lines of human EC cells have been characterized to extend this approach to human embryogenesis (3), while the isolation of embryonic stem (ES) cells (4, 5) has opened up important new approaches for the study of gene function during embryonic and fetal development. Many general aspects concerning the use of such cells were extensively covered in an earlier volume in this series (6), and the interested reader is encouraged to consult the relevant chapters of that volume for more comprehensive information. In the present chapter we shall repeat only the basic details necessary for working with these cells, and aim to supplement that earlier work with information more directly relevant to neurobiologists, and with newer techniques, such as homologous recombination and gene 'knock-out' experiments, which were not covered in detail previously.

2. Teratocarcinomas as tools in embryology

2.1 Origins of teratocarcinomas and their relationship to the early embryo

Teratocarcinomas are histologically complex tumours that arise most frequently, but not exclusively, in the gonads, and belong to a class of neoplasms known as germ cell tumours (GCT). Ovarian dermoid cysts, which develop from parthenogenetically activated oocytes that begin embryonic development but then become disorganized, are the most common manifestation of

GCT. These tumours may contain many embryonic and fetal tissues and grow to a large size, but they are generally benign. By contrast, testicular GCT, which arise from transformation of primitive germ cells prior to meiosis, are usually malignant. Such tumours may contain one or more elements that resemble extraembryonic tissues of the trophoblast or yolk sac (choriocarcinoma and yolk sac carcinoma cells, respectively) or various embryonic tissues (teratoma). Most testicular GCT also contain cells known as embryonal carcinoma (EC) cells which are the malignant stem cells from which the other elements arise by differentiation; in many instances, but not all, the differentiated derivatives of EC cells have a limited proliferative capacity and have lost their malignant properties. The term teratocarcinoma is usually reserved for GCT that contain both embryonic tissues (teratoma) *and* EC cells. A more detailed summary of GCT pathology is provided in the earlier volume in this series (7).

The discovery that mice of the 129 strain develop spontaneous testicular teratocarcinomas (8), and that these tumours can be produced in other strains from ectopically transplanted embryos (for example, to the kidney capsule) (2), allowed the experimental study of teratocarcinomas in the laboratory mouse. Cell lines composed of EC cells were isolated and shown to be pluripotent: after single-cell cloning such EC cells are capable of extensive differentiation into the many cell types found in teratocarcinomas (9). Further studies showed that, after injection into the blastocoel cavity of a developing embryo, pluripotent EC cells can participate in normal embryogenesis with their differentiated progeny contributing to the tissues of the chimeric mouse that develops (10, 11). The consensus from many studies was that murine EC cells most closely approximate cells of the primitive ectoderm of the early embryo, and that their differentiation, which can also be induced *in vitro*, provides a window on to the processes of early embryonic cell differentiation. By analogy, human EC cells should be able to provide information about cells of the early human embryo.

Neuronal differentiation by murine and human EC cells *in vitro* has often been reported (see, for example, refs 12–15) and, in a few cases, both neuronal and astroglial differentiation has been clearly demonstrated (16). The evidence for the identity of the derivative cells includes the expression of specific markers, such as neurofilament and glial filament proteins, and functional tests of electrophysiological properties (17, 18). From the perspective of the neurobiologist, the probable resemblance of EC cells to the primitive ectoderm implies that, in contrast to other tumour models for neurogenesis, teratocarcinomas provide a model for the very earliest stages of neural differentiation, and for exploring the molecular processes whereby pluripotent embryonic cells choose between neural and non-neural pathways of differentiation.

Despite their resemblance to embryonic cells, it is important to remember that EC cells are derived from tumours. As such, they are frequently karyotypically abnormal. Further, since the process of differentiation commonly leads to cells with a limited proliferative potential, there is a strong selective pressure in tumours for the growth of EC cells that have lost or reduced their potential for differentiation. Consequently, many EC cell lines lack the capacity for differentiation (they are said to be 'nullipotent') while few, if any, possess the full developmental repertoire of the primitive ectoderm. Recently, however, techniques have been devised for isolating and maintaining fully pluripotent stem cells (ES cells) directly from the early embryo (4, 5). These ES cells resemble EC cells in many ways, including their expression of the same markers and their ability to form teratocarcinomas when injected into syngeneic mice. However, they are karyotypically normal and are able to contribute at a much higher frequency to all tissues, including the germ cells, of chimeric mice that result after their injection into blastocysts (19). Details concerning the derivation and maintenance of ES cell lines are provided by Robertson (20).

2.2 Genetic manipulation using ES cells

ES cells retain their pluripotentiality after prolonged growth and genetic manipulation *in vitro*, thus providing a unique and powerful link between tissue culture and the mouse genome. The potential for the determination of specific gene function—for example, in relation to development of the nervous system—is tremendous. With recent refinements in DNA construct design and transfection techniques, coupled with improved methods of mutant selection, it is now possible to isolate ES cell clones carrying precisely targeted mutant genes and to create injection chimeras that transmit the mutant allele to their progeny. This is achieved by introducing gene targeting DNA constructs into ES cells and selecting for the rare clones which have undergone homologous recombination between the construct and an endogenous locus (reviewed in refs 21–23). To date, there have been more than a dozen reports of mutant strains of mice generated by such gene 'knock-out' experiments in which the activity of a putative developmentally important gene has been eliminated. Some of these mutants have neurologic defects, both dramatic (24, 25) and subtle (26), helping to elucidate the functions of the genes in question during normal nervous system development. Rossant has recently reviewed this and other means of manipulating the mouse genome from the perspective of neurobiology (27).

3. Cell lines of relevance to neurobiology: their maintenance and differentiation

3.1 Introduction

A large number of murine EC and ES lines have been derived during the past two decades, and it seems unlikely that researchers with a primary interest

Table 1. EC cell lines useful for studies in neurobiology

Cell-line Sublines	Species	Origin	Induction of differentiation	Refs
P19 P19S1801A1 (01A1)	Mouse C3H/HeHa	Embryo-derived tumour	Chemical RA $>10^{-7}$ M	16, 29–31
PCC7-S PCC7Saza^{R1} -1009	Mouse Rl 129XB6	Testicular tumour	Spontaneous Chemical (RA)	32–34
PCC3 PCC3/A/1	Mouse, 129Sv Sl^J/+	Testicular tumour	Spontaneous	12, 35
PCC4 PCC4F PCC4aza^{R1}	Mouse 129Sv Sl^J/+	Testicular tumour OTT6050	Chemical (RA)	35–37
C17-S$_1$ 1003	Mouse C3H/He	Embryo-derived tumour	Spontaneous in serum-free medium	32, 38–40
TERA-2 NTERA-2 cl.D1 (NT2/D1) TERA-2 cl.13	Human	Testicular tumour; lung metastasis	Chemical (RA)	14, 41–43

in neurobiology will wish to invest time in deriving new lines. Those who do should consult the chapters by Damjanov *et al.* (28) and Robertson (20) in the earlier volume in this series.

Neuroepithelial differentiation is a common pathway followed by pluripotent EC cells, and it is not possible to review all of the available EC lines here. Rather, we shall focus on a few lines (*Table 1*; *Figure 1*) that are widely available and have been used in studies of developmental neurobiology. The techniques for handling these lines, which are discussed in detail below, are illustrative of approaches that are generally applicable to other EC cell lines.

3.2 EC cell lines

3.2.1 General points

i. Growth and maintenance

Many of the widely used EC cells have been adapted to grow without feeder layers. This is not true of ES cells which generally require a feeder layer, although they can sometimes be grown without feeders if the growth factor LIF/DIA is present (44); if it is necessary to use one of the EC lines that require feeders, the protocols described below regarding ES cells may be used. In *Protocol 1* we have described a generic protocol for growing EC cells, and it is one that we commonly use. However, either because of the

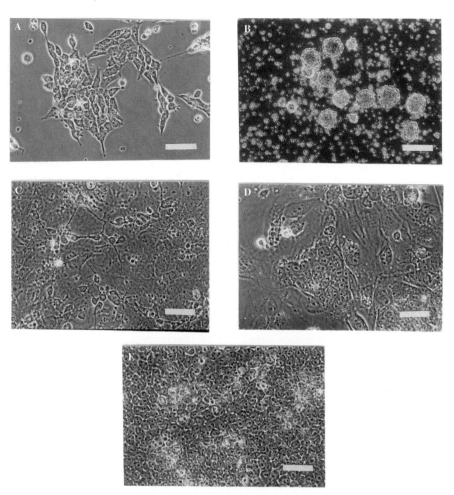

Figure 1. The morphology of EC and ES cells. **A**: Mouse EC cell line PCC7S (clone 1009) growing in monolayer culture. **B**: Embryoid bodies formed by PCC7S cells growing in suspension, 4 days after seeding. **C**: Neurons and other differentiated cells growing from PCC7S embryoid bodies 7 days after reseeding in a tissue culture in the presence of 10^{-6} M RA. **D**: CCE ES cells growing on STO feeders; note the difference in morphology between the tight cluster of ES cells and the surrounding STO cells. **E**: Confluent culture of NT2/D1 human EC cells; note the similarity in morphology to the PCC7S mouse EC cells (**A**) and the CCE mouse ES cells (**D**). Bar = 50 μm panels **A, C, D, E**; Bar = 200 μm panel **B**.

particular needs of specific cell lines, or because of the local practices of individual laboratories, variations of this technique are often described. For example, Dulbecco's modified Eagle's medium (DME) (high glucose formulation) has been used historically by many investigators, but others have preferred the Alpha modification of Eagle's minimal essential medium. In our experience the single most significant variation in culture condition requirements results from the effect of cell density upon differentiation of pluripotent cells: Some lines, especially the human lines with which we have worked, should be grown at high cell densities ($>7 \times 10^4$ cells per cm^2) to maintain their undifferentiated state, while other lines are better maintained at lower cell densities. Significant variations that apply to the cell lines listed in *Table 1* are discussed below.

Standard tissue culture plastic dishes and flasks from any of the major suppliers are generally satisfactory for culturing EC cells. Typically, EC cells grow in clusters which often grow upwards rather than outwards, presumably due to preferential adhesion of the cells to one another than to the substrate. If this is a problem, attachment and spreading of the cells can be enhanced by gelatinizing the culture dish. To do this, cover the growing surface of the dish with a sterile 0.1% solution of gelatin in PBS, and leave for about an hour at 4°C. Then aspirate the gelatin and seed the cells as described in *Protocol 1*. A sterile stock solution of 10% gelatin in water may be prepared by autoclaving and dividing it into convenient small aliquots before it gels. When required, melt an aliquot in a water bath and dilute 1:100 in sterile PBS; the resulting 0.1% solution does not gel and may also be stored for some time.

Whichever culture medium is used, it can be purchased in powder or liquid form from any of the major suppliers; no special precautions are necessary beyond those required by good cell culture practice in general. It is best to avoid the use of antibiotics if possible, since they are unnecessary and their use can mask sloppy technique! However, if required, common antibiotics such as penicillin, streptomycin, or gentamycin may be used without any obvious side-effects. As in the case of cell culture generally, cultures should be regularly tested for mycoplasma contamination.

The medium is usually supplemented by FCS, although some investigators use a mixture of FCS and calf serum. Before use we inactivate the serum by heating to 56°C in a water bath for 30 min. Different lots of serum from a reputable company should be screened in order to select a batch that is optimal for growth *and* differentiation. In our experience, lot to lot variation is often greater with respect to the promotion of differentiation than with respect to maintenance of the EC cells themselves.

Protocol 1. A generic protocol[a] for the harvesting and passaging of EC cell lines that do not require feeders

1. From a 75-cm^2 tissue culture flask of cells to be harvested, aspirate the medium and rinse with about 5 ml Dulbecco's phosphate-buffered saline (PBS) lacking calcium and magnesium.

2. Add 1 ml of trypsin–EDTA (0.25% trypsin, 1 mM EDTA in calcium- and magnesium-free PBS), warmed to 37°C. Depending on the cell line, trypsin

may be omitted; in the latter case 2% chicken serum can help to maintain cell viability.

3. When most of the cells have begun to detach (2–4 min), tap the flask sharply two or three times with the palm of the hand to dislodge the remaining cells.

4. Add 9 ml fresh culture medium containing 10% FCS, pipette up and down two or three times to break up clumps (but not too vigorously!) and transfer to a sterile tube. (It is not usually necessary to wash the cells for routine passaging, provided sufficient fresh medium containing FCS has been added to inactivate the trypsin:EDTA; if washing is desired, pellet the cells by centrifuging at 200–400 g for 3–4 min.)

5. Count the cells with a haemocytometer. (Generally, cell viability is greater than 90%. To check this, dilute a small aliquot of the cell suspension with an equal volume of a 0.15% solution of erythrosine B in PBS, before placing in the hemocytometer chamber; dead cells stain red, live cells do not stain. Remember to adjust for the dilution with erythrosine B when calculating cell numbers.)

6. Reseed in a fresh 75-cm^2 tissue culture flask by transferring a volume of the cell suspension containing 10^6 cells; adjust the cell number proportionately if a culture vessel of a different growing area is used. Add 15 ml fresh medium (DME, high glucose formulation, supplemented with 10% FCS, is commonly used) and culture at 37°C under a humidified atmosphere of 5% CO_2 in air.

[a] Variations on this theme are used by different laboratories; variations used for particular cell lines are discussed in the text.

ii. Differentiation

The two broad approaches to inducing EC cell differentiation *in vitro* are either to alter the growth conditions to promote spontaneous differentiation, or to add a chemical inducing agent. Combinations of altered growth conditions and chemical inducer may also be used.

Spontaneous differentiation. The most generally used method for promoting spontaneous differentiation of murine EC cells is to increase the cell density. This can be achieved in some cases (for example, PCC3) merely by allowing cultures to grow to confluence and then maintaining them in that condition for several days or weeks, feeding at regular intervals. The disadvantage of this approach is that it often results in a large amount of cell death. The alternative is to force the cells to form clusters by growing them in suspension in bacterial Petri dishes (*Protocol 2*; *Figure 1B*). Since the plastic in these has not been modified for cell culture, the EC cells do not attach to the substrate but adhere to one another forming floating clusters. These differentiate to form 'embryoid bodies', in which an inner core of EC cells is surrounded by a rind of cells that resemble the primitive endoderm (45). More extensive differentiation may occur if embryoid bodies are held in suspension for some days. Subsequently, if these embryoid bodies are transferred to a tissue culture surface they attach and differentiated cells grow out from them.

Protocol 2. Induction of EC cell differentiation by forming aggregates in suspension culture

1. Prepare a single-cell suspension.
2. Plate at 10^5 cells per millilitre in a bacterial grade Petri dish, in the usual growth medium. For some cells it may be important to include retinoic acid (RA); see discussion of individual cell lines in the text.
3. Culture for 4–5 days at 37°C. Feed with fresh medium after 3–4 days. The cells should form floating clusters (*Figure 1B*); if they show marked adherence to the surface of the dish, repeat from step 1, but first coat the dish with a thin layer of sterile 0.2% agarose.
4. After 4–5 days, carefully transfer the floating clusters to a tissue culture dish with fresh medium and continue culturing. The clusters attach and differentiated cells grow out. Sometimes the addition of RA during the outgrowth phase is helpful.

In contrast to murine EC cells, the human EC cells we have studied tend to be inhibited from differentiation when held at a high density. In several cases, some degree of differentiation may be induced by culture at low density, but so far we have not seen evidence of neuroepithelial differentiation induced by this growth condition.

Chemically induced differentiation. By far the most widely used agent to induce EC cell differentiation is all-trans-retinoic acid (RA). This can be obtained from several chemical companies; we generally buy it from Eastman Kodak or Sigma. As it is light-sensitive and subject to oxidation, long-term storage of the powder should be under nitrogen in the dark. Stock solutions (10^{-2} M) are prepared by dissolving in DMSO, at a concentration of 3 mg per millilitre; aliquots of such stock solutions can be kept for several months at −70°C. RA is added to the culture medium by diluting an appropriate volume of the stock solution directly in the medium. At the highest concentration of RA commonly used (10^{-5} M), this also results in a DMSO concentration of 0.1%. This concentration of DMSO is not usually a problem, but at higher concentrations DMSO may itself induce differentiation, depending upon the cell line used. Therefore, a control culture including an appropriate concentration of DMSO but no RA should be included in the experimental protocol. If DMSO is a problem, ethanol can be substituted as the solvent for preparing the stock solution.

3.2.2 Selected EC cell lines (*Table 1*)

i. P19

P19 is a murine EC cell with a 40, XY karyotype; various clones have been described, notably P19S18O1A1 which is both ouabain- and 6-thioguanine-resistant. The undifferentiated EC cells are maintained in the Alpha modification of Eagle's minimal essential medium, but without nucleosides or deoxyribonucleosides, and supplemented with 2.5% FCS and 7.5% calf serum.

Differentiation may be induced by plating cells in suspension culture in the presence of 10^{-5} M to 5×10^{-7} M RA, as described in *Protocol 2*. RA is also included in the medium once the aggregates are placed in tissue culture dishes. Under these conditions, most of the EC cells differentiate and neurons and glial cells are prominent among their derivatives (16).

P19 is a pluripotent EC cell line and its direction of differentiation can be modulated by the culture conditions. In the absence of RA, the aggregates in suspension culture form simple embryoid bodies consisting of EC cells surrounded by endoderm; neurons do not grow out when these are allowed to attach. Neither do neurons form if differentiation is induced directly in monolayer culture by addition of RA without first forming aggregates in suspension. Further, concentrations of RA lower than 5×10^{-7} M result in muscle differentiation rather than neuronal differentiation; 10^{-8} M induces skeletal muscle formation whereas 10^{-9} M induces predominantly cardiac muscle (30). DMSO (0.5–1.0%) also induces differentiation, but into muscle, not neurons; a combination of DMSO and RA ($> 5 \times 10^{-7}$ M) results in the differentiation characteristic of RA alone—i.e. neuronal, but no muscle differentiation (31).

ii. PCC7S/1009

PCC7-S and its azaguanine resistant clone PCC7-S AzaR1 clone 1009 (usually abbreviated to simply 1009) may be maintained as described in *Protocol 1* (*Figure 1A*). These EC cells can be induced to undergo spontaneous differentiation by the formation of aggregates in suspension culture as described in *Protocol 2* (*Figure 1B* and *C*). Cholinergic neurons are prominent among the cells that grow out from the aggregates when allowed to reattach to a tissue culture surface (33) preferably in the presence of retinoic acid (10^{-6} M). In this case neurite outgrowth is enhanced by overlaying the attached aggregates with medium containing 1.3% carboxymethylcellulose. Further, the addition of 10^{-5} M cytosine arabinoside inhibits the outgrowth of non-neuronal cells.

iii. PCC3

These EC cells can be maintained as in *Protocol 1* and induced to undergo spontaneous differentiation by either maintenance of a monolayer culture at high cell density or by the formation of aggregates described in *Protocol 2*.

The differentiated cultures obtained from these cells are quite heterogeneous, but do include neurons.

iv. PCC4

PCC4 EC cells, maintained as in *Protocol 1*, do not generally undergo spontaneous differentiation, although some sublines will give endoderm-like derivative after culturing in suspension. Differentiation can be induced with retinoic acid (36), and neurons are found among the derived cells (17, 37). However, RA is toxic to some sublines of PCC4 cells. This can be overcome by a process of adaptation and selection, by growing cultures in low concentrations of RA ($< 10^{-8}$ M) for several passages, slowly increasing the concentration until a resistant line is derived (36). Such resistant lines can be induced to differentiate by growth in suspension in the presence of RA.

v. 1003

The EC line C17-S1 clone 1003, described by Darmon *et al.* (38), is unusual in that it can be induced to undergo spontaneous neuronal differentiation by growth in serum-free medium (SFM). Presumably FCS contains factors that inhibit the neuronal differentiation of this cell. Clone 1003 EC cells can be maintained by the procedure in *Protocol 1*. To induce differentiation, a culture of these cells is gently washed with SFM (*Table 2*), and then harvested directly into SFM by flushing more vigorously. The suspended cells which detach in small clumps are reseeded at about 10^4 cells per cm^2 into tissue culture dishes in SFM F4 (*Table 2*). Islands of neuron-like cells extending long axons appear after 6–7 days (38). Gelatinizing the dishes prior to seeding cannot substitute for fibronectin in the SFM formulation, but can improve the health of the subsequent cultures.

vi. TERA-2

TERA-2 is a human EC cell line and a number of clones have been derived, NTERA-2 cl.D1 and TERA-2 cl.13 being the most commonly used. Unlike

Table 2. Serum-free medium, F4, for promoting differentiation of 1003 EC cells (38)

Basal medium:	50% DME:50% Ham's F12[a]
Additives:	
Insulin[b]	5 μg/ml
Transferrin	10 μg/ml
Sodium selenite	3 × 10^{-8} M
Fibronectin	5 μg/ml

[a] DME:F12 mix is now sold by several companies.
[b] To prepare a stock solution of insulin at 1 mg/ml: dissolve 10 mg bovine insulin in 1 ml 0.1 M HCl; when dissolved, dilute with 9 ml PBS, aliquot, and store frozen.

most murine EC cells these cells tend to differentiate if grown at low cell densities or even if maintained over the long run by trypsinization (42). To avoid this problem, stock cultures should be passaged by scraping the cells, rather than by using trypsin or EDTA, and reseeding at cell densities greater than 7×10^4 per cm^2. Typically, a confluent 75-cm^2 flask of NTERA-2 cl.D1 (*Figure 1E*) yields about 2×10^7 cells; since cells cannot be counted accurately after scraping because they are in clumps, such a confluent culture should be split in a 1:3–1:4 ratio to achieve the correct cell density. An additional problem with these cells is that they acidify the medium fairly quickly, so that it is necessary to use rather more medium (20–25 ml per 75-cm^2 flask) than with other cell types. Cultures of NTERA-2 cl.D1 cells maintained in this way should be passaged every 3–4 days.

To induce differentiation of NTERA-2 cl. D1 EC cells, harvest a stock culture using trypsin:EDTA, and reseed at a density of 10^6 cells per 75-cm^2 flask in medium containing 10^{-5} M RA. Such cultures should be re-fed at weekly intervals. Under these conditions most cells loose their EC phenotype (for example, as judged by expression of surface antigen markers) within 7–10 days, and cells expressing other markers appear. Nevertheless, the cells continue to proliferate, although initially at about half the rate of the EC cells, and the cultures achieve a density of 2–3×10^7 cells after 14–21 days. Neurons can be found in the cultures during the second week. However, as they are often obscured by the high density of other cell types, they are most easily visualized by staining with antibodies to neurofilament proteins (46). Selection of the FCS used in these experiments is important in optimizing the formation of neurons.

To obtain cultures in which the neurons are more prominent, trypsinize such a differentiated culture after 3–4 weeks in RA, and reseed the cells at a 1:4 split ratio in the absence of RA. It may be helpful to reduce the serum concentration to 5%. The resulting cultures grow to confluence and then enter a quiescent stage which can often be maintained for several months if fed but left otherwise undisturbed. Clusters of neurons with interconnecting bundles of axons form on top of the underlying monolayer of other cell types during the 2–3 weeks after the cultures reach confluence (*Figure 2*). The neurons in these cultures have low sodium channel densities (1–2 channels per μm^2), giving them electrophysiological properties consistent with those of some embryonic neurons (18). Frequently, cells that exhibit a neuronal morphology and express neuronal markers, such as neurofilament proteins, account for about 5% of the cells in such cultures, although a rather higher proportion of cells may express neurofilament proteins but not resemble neurons morphologically. The proportion of neurons can be enhanced by incorporating 10^{-5} M cytosine arabinoside in the medium for 3–4 days after the RA-induced cells are reseeded.

Differentiation of NTERA-2 cl.D1 cells may alo be induced by other agents, notably hexamethylene bisacetamide (HMBA) (47, 48). In this case,

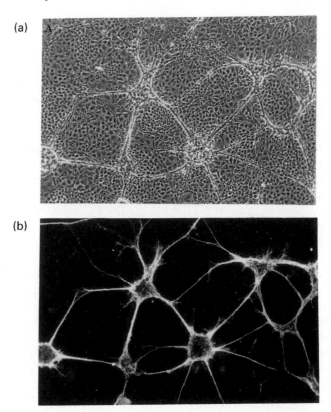

Figure 2. Neuronal differentiation from NT2/D1 human EC cells. After culture for 3 weeks in 10^{-5} M RA, the cells were reseeded in the absence of RA. These neuron clusters overlying other differentiated cells developed over a subsequent 2–3 week period; they were fixed and stained by immunofluorescence for neurofilament expression. **A**: phase contrast; **B**: immunofluorescence staining for the 200 kDa neurofilament protein.

seed the cells as for RA induction, but replace the RA with 3 mM HMBA in the medium, diluted from a 0.3 M stock solution in water. The kinetics of differentiation are similar to RA-induced differentiation, but the cells formed exhibit a quite different phenotype from those induced by RA, and few if any neurons appear.

Unfortunately, apart from the neurons, the identity of the differentiated cells induced by either RA or HMBA is unknown. In neither case do they appear to correspond to extraembryonic cell types such as trophoblast or endoderm, or to somatic cell types such as muscle. Their relationship to other neuroepithelial cell types is unclear; especially in the RA-induced cultures, many cells express various ganglioside antigens, and also NCAM, but evidence of glial differentiation is lacking.

3.3 ES cell lines

3.3.1 General points

The culture of ES cells is very similar to the culture of feeder-dependent EC cell lines. The pluripotentiality of mouse ES cells is maintained in culture by a factor called leukemia inhibitory factor or differentiation inhibiting activity (LIF/DIA), present either in the culture medium or in the extracellular matrix of certain feeder cells (44, 49). Traditionally, STO cells have been the feeder cells of choice for all studies involving EC and ES cells. They may be obtained from the ATCC. Although ES cells can remain pluripotential after extended periods *in vitro*, it is prudent to freeze aliquots of cell lines or clones early to avoid overly prolonged passage, with the consequent possibility of selecting for subpopulations of cells with reduced pluripotentiality. Details of ES cell culture are extensively discussed in a previous volume in this series (20), but will be briefly reviewed here.

The following established ES cell lines have commonly been used in gene targeting experiments and have successfully contributed to the germ-line in chimeras: D3 (50); CCE (51); E14 (52); CC1.2 (19); AB-1 (24); AB-2 (53). All of these lines have been derived from strain 129 mice and are maintained as outlined below. They all give rise to chimeras with extensive ES-derived contribution (50–100%) with efficient transmission of the ES-derived genome through the germ-line. The successful maintenance of pluripotentiality during extended *in vitro* manipulation seems to depend less on which cell line is used, and more on how carefully the ES cells (and feeder cells) are maintained. However, cell lines from one laboratory (AB-1, AB-2) have exhibited a truly remarkable contribution to chimeras (24, 53).

Although a normal karyotype and an undifferentiated morphology (*Figure 1*) are essential characteristics as a measure of the pluripotentiality of ES cell cultures, these qualities alone do not assure that culture conditions are sufficient. The best way to assess adequacy of cultures in a particular laboratory is first to grow cultures for one or two weeks and then to perform blastocyst injections with these cells. If the ES-derived contribution to the chimeras is low, it may be concluded that the cultures conditions are suboptimal. Adequate culture conditions should result in survival of about 50% of injected blastocysts, with about 50% of these being chimeric and with the ES cell contribution to most chimeras being greater than half (remember '50:50:50').

i. Growth and maintenance of ES cells

The growth of ES cells and maintenance of a pluripotential phenotype is relatively straightforward if care is taken to assure that culture conditions are adequate. They grow very rapidly and are thus easily expanded to the large numbers of cells necessary for gene targeting experiments.

LIF/DIA can be supplied by culturing the ES cells on feeder monolayers of specific cell lines such as the permanent mouse embryonic fibroblast line, STO, or primary embryonic feeder cells. Alternatively, ES cells may be cultured with LIF/DIA present in the supernatant either as conditioned medium from specific cell lines (for example, 5637 human bladder carcinoma or Buffalo rat liver cell lines), or in recombinant form ('ESGRO', Gibco-BRL). All of the various sources of LIF/DIA have been used to maintain pluripotential ES cells which contributed to the germ-line in chimeras (discussed in refs 44, 49). There are advantages and drawbacks to each culture system, and in general, most investigators tend to use one particular culture methodology.

The use of STO feeder-cell monolayers is straightforward and has the advantage that the feeder cells perform other functions besides the production of LIF/DIA. The feeder cell monolayer provides a better substrate for ES cell attachment thereby increasing plating efficiency (and thus increasing the total number of G418-resistant colonies in selection experiments). Culture with feeder layers may also be less affected by variabilities in culture conditions. The chief disadvantages are that additional time and resources are required for production of feeders, and carryover feeders might confuse cell counts and the microscopic selection of cells for blastocyst injections.

Primary embryonic feeder cells support ES cell growth well but they must be periodically rederived due to their limited life-span, and any G418 selections of ES cells must be performed on embryonic feeder layers derived from mice bearing a transgenic neomycin-resistance gene. Sublines of G418-resistant STO cells are easily generated by transfection with a neo^r plasmid (23).

Culture on gelatinized plates in the presence of exogenous LIF/DIA seems more sensitive to variabilities in the media components and has a lower plating efficiency, but we have found that G418 selections can be performed more rapidly than on feeder layers thus minimizing the time of exposure to the selection agents. With prior testing of reagents to assure a lack of toxicity to ES cells, this method of culture is somewhat simpler than coculture with feeder.

Protocols 3 and *4* outline the culture conditions for STO feeder cells and mouse ES cell lines. Tissue culture plates and flasks should be gelatin-coated (as above, Section 3.2.1) and can be stored dry, at 4°C for long periods. Stock cultures of STO cells are straightforward to maintain but are of such importance to the successful growth of ES cells that their culture warrants particular care. They are maintained in DME + supplements by splitting 1:10 every 3 days. Never let the cultures remain confluent for extended periods, since this selects for cells which have decreased contact inhibition and are poor feeder cells.

Embryonic stem cells are grown by plating on mitotically inactivated STO feeder layers (*Protocol 3*) in DME (high glucose formulation) + 10% FCS (heat-inactivated), supplemented with 1× non-essential amino acids, 1 mM glutamine, and 0.10 mM β-mercaptoethanol (hereafter referred to as 'DME + supplements'). Acceptable lots of FCS are chosen after comparison of their affect on ES cell colony morphology and plating efficiency at clonal densities. DME + 10% FCS can be stored in the dark for weeks at 4°C, but supplements should be added just prior to use. The β-mercaptoethanol should be made weekly as a 1000× stock (add 70 μl of β-mercaptoethanol to 10 ml PBS and filter sterilize).

Protocol 3. STO feeder layers

1. Remove the medium from a confluent culture of STO cells, replace with fresh medium containing 10 μg/ml mitomycin C[a] and incubate at 37°C for 2–3 h. Alternatively, the cells may be mitotically inactivated by irradiating the culture with 3–5000 rads, if a suitable γ-irradiator is available.
2. Rinse the flasks three times with PBS, harvest by trypsinization, and pool the cell suspensions from all the flasks.
3. Count viable cells (each confluent 75-cm^2 flask should yield about 8–10 × 10^6 cells). Determine the volume of cell suspension needed to plate 5 × 10^4 cells per cm^2 (i.e. 3.7 × 10^6 per 75-cm^2 flask; 3.1 × 10^6 per 10-cm Petri dish; and 9 × 10^4 per well of a 24-well plate).
4. For the best results, allow feeders to attach for several hours to form a monolayer before plating the ES cells, although feeders and ES cells can be plated simultaneously if necessary.
5. Feeder cell monolayers may be kept in the tissue culture incubator and used up to a week after plating. Feed such feeder cultures by the addition of fresh medium after several days.

[a] Mitomycin-C (Sigma) should be made fresh every 2 weeks and stored at 4°C, protected from light.

Protocol 4. Routine culture of ES cells

ES cells are grown rather densely and are split every 2–3 days at 1:10 split ratio.

1. Harvest ES cell cultures from 75-cm^2 flask by trypsination (as above, *Protocol 1*).
2. Inactivate trypsin by addition of several millilitres of fresh DME + supplements and pipette gently several times. ES cells adhere to one another very tightly, so it is good practice to leave the cell suspension in the flask and examine microscopically to assure a single cell suspension has been achieved.[a]
3. Add 1/10 of this volume to each 75-cm^2 flask containing a fresh feeder cell monolayer (*Protocol 3*) and 15 ml of pre-warmed DME + supplements.

Protocol 4. *Continued*

4. Adjust the split ratio from 1:10 if there seem to be too many or too few colonies in the subsequent cultures.

---

a If large clumps are present, rather than try to pipette more vigorously to disrupt the clumps, simply pipette the suspension into a centrifuge tube, and pellet the cells by centrifugation (3–500 g, 5 min), aspirate medium, resuspend the pellet in 1 ml of trypsin–EDTA and incubate another 5 min at 37 °C.

---

ii. *Differentiation of ES cells*

Similar to EC cell lines, ES cells will differentiate spontaneously when cultured in the absence of LIF/DIA or they can be induced to differentiate by chemical agents as mentioned above. These techniques are discussed extensively in the previous volume (6). The unique aspect of ES cells is their *in vivo* differentiation and contribution to all the tissues in chimeras.

4. Stem cell markers and monitoring differentiation

Embryonal carcinoma cells are inherently unstable because of their capacity for differentiation. Thus, the proportion of EC cells in a culture can be seriously diminished if poor attention to growth conditions leads to excessive spontaneous differentiation. Consequently, it is important to monitor stock EC cultures continuously and to be alert for appearance of significant numbers of differentiated cells that might interfere with subsequent experiments.

Generally, EC cells exhibit a characteristic morphology that distinguishes them from their differentiated derivatives (*Figure 1*), and phase microscopy is convenient for quickly checking the state of cultures. Typically, EC cells grow in tight clusters of small cells with a high nuclear:cytoplasm ratio, and with few prominent nucleoli. By contrast, the differentiated cells that commonly arise spontaneously in the early stages of differentiation are markedly bigger and flatter.

A more objective and quantitative approach to monitoring cultures is to assay expression of certain specific markers. Several cell surface antigens are characteristically expressed by EC cells (*Table 3*), and their expression is conveniently assayed by immunofluorescence (*Protocol 5*). This type of assay is also useful in monitoring the progress of differentiation induced by particular experimental protocols. In that case it is also useful to assay the appearance of antigens characteristic of differentiated cells, as well as the loss of EC markers.

Apart from markers that might be of specific interest to neurobiologists (such as those expressed by neurons), several surface antigens in addition to those listed in *Table 3* have often been used in studies of teratocarcinomas.

Table 3. Surface antigen markers of EC cells (Andrews 1988)

Antibody			Antigen	Expression in teratocarcinomas				
				Mouse		Human		
Name	Class	Species		EC	Diff.[a]	EC	Diff.[a]	Ref.
MC480	IgM	Mouse	SSEA-1	+++	–	–	+	54
MC631	IgM	Rat	SSEA-3	–	+	+++	–	55, 56
MC813-70	IgG$_3$	Mouse	SSEA-4	–	+	+++	–	57
TRA-1-60	IgM	Mouse	TRA-1-60	–	–	+++	–	58
TRA-1-81	IgM	Mouse	TRA-1-81	–	–	+++	–	58

[a] Note the distinction between murine and human EC cells; the differentiated cells do not necessarily or uniformly express the antigens shown, but appearance of these antigens together with disappearance of EC marker antigens can provide a useful means of monitoring the progress of differentiation.

For example, Class 1 MHC antigens (H-2) often appear during murine EC cell differentiation, although the expression is not always easily detectable. The Thy-1 antigen also may appear during murine EC cell differentiation (remember that most murine EC cells are derived from mouse strains carrying the Thy-1.2 allele). Neither of these markers are so useful for human EC cell differentiation, since they are expressed by the undifferentiated EC cells, albeit weakly and variably in the case of MHC antigens (3). On the other hand several ganglioside-associated glycolipid antigens appear on EC cell differentiation. One that we have used extensively in the case of NTERA-2 differentiation is the A2B5 antibody (59), which detects the ganglioside GT$_3$ in retinoic-acid-induced cells (60). Antibodies recognizing these antigens, and also the antibodies listed in *Table 3*, are available from the American Type Culture Collection or from the Developmental Studies Hybridoma Bank.

Protocol 5. Immunofluorescence assays for the expression of cell surface antigens

1. From the culture to be tested, harvest the cells using trypsin–EDTA, wash once in fresh medium (see *Protocol 1*) and suspend to 10^7 per millilitre in PBS containing 5% FCS.
2. Mix 50 μl of the cell suspension and 50 μl of the monoclonal antibody to the antigen to be assayed, diluted as appropriate,[a] in a well of a round-bottomed 96-well plate; seal the plate with adhesive sealing tape and incubate at 4°C for 1 h, preferably on a plate shaker.
3. Centrifuge the plate at 400 g for 2–3 min to pellet the cells and aspirate the supernatant.

Protocol 5. *Continued*

4. Add 100 μl of cold-wash buffer (PBS plus 5% FCS and 0.1% sodium azide) to the well and repeat step 3.
5. Repeat step 4 twice. After the last centrifugation aspirate the supernatant and go to step 6.
6. Add 50 μl of a predetermined dilution of a fluorescence-conjugated antibody with a specificity appropriate for the first antibody. For example, if the first antibody were a mouse IgM, the second antibody should be a goat anti-mouse IgM, and conjugated to a fluorochrome such as fluorescein. Good quality commercial preparations are available from a number of companies and should be titrated to determine an appropriate dilution. For most purposes the cheaper preparations made from only partially purified IgG are sufficient.
7. Seal the plate and incubate as in step 2.
8. Wash the cell three times as in steps 3–5.
9. After the last wash suspend the cells in 50 μl 50% glycerol in PBS and mount under a coverslip on a glass slide for examination under a microscopy equipped for fluorescence studies. Alternatively, suspend the cells to 10^6 per millilitre in PBS for flow cytofluorimetry if that equipment is available.

[a] The appropriate dilution of the first antibody must be determined in each case by titration. However, as a rule of thumb, antibodies available as culture supernatants should be generally used neat, while ascites preparations are usually good at dilutions of 1:100, and often down to 1:1000 or more.

5. Production of mutant mice by homologous recombination in ES cells

5.1 General considerations

In contrast to yeast, exogenous DNA usually integrates randomly in mammalian cells and only rarely will recombination take place with homologous regions of the genome. The major challenge in gene targeting by homologous recombination is in the isolation of targeted cells from among the enormous number of cells which either have not integrated the transfected DNA, or which have done so at random sites. A number of enrichment schemes and screening techniques have been successfully employed (reviewed in refs 21–23).

Herein we will briefly discuss:

(a) the design of a targeting construct to maximize the efficiency with which one can isolate clones of ES cells in which the gene of interest has undergone homologous recombination with the transfected DNA;
(b) the means by which one introduces the DNA into ES cells; and
(c) the means by which one can select and characterize the transfected cells.

A brief discussion of the techniques for transfer of ES cells into blastocysts for the generation of chimeric mice will be presented, but this has also been extensively covered in a previous volume (61).

5.2 Design of gene targeting constructs

5.2.1 General considerations

A targeting construct consists of a plasmid containing a cloned region of genomic DNA from the gene of interest (usually spanning from a few kilobases to over 15 kb) into which some alteration has been introduced. Typically, the neomycin-resistance gene (neo^r) is inserted into an exon of the gene both to disrupt the normal coding sequence and to act as a positive selection element to allow for survival of cells which have integrated the construct DNA. Large-scale screening of neomycin-resistant colonies can be performed, but the frequency of targeted clones resulting from homologous recombination is usually very low (often 1 homologous recombinant per about 1000 G418-resistant colonies). However, targeting constructs can be created which greatly enrich for targeted clones from among surviving colonies. The appropriate enrichment scheme to use depends upon whether or not the gene of interest is normally expressed in the ES cells; many genes affecting development of the nervous system are probably not expressed in ES cells.

5.2.2 Transcriptionally active genes

Genes which are expressed in ES cells can be very efficiently targeted for homologous recombination by the use of a targeting construct containing a promoterless neo^r gene (62). In this case, only those random sites of integration which are in correct orientation near an active promoter will express the neo^r gene. Homologous recombination at the targeted locus will cause expression of the neo^r gene so that the surviving G418-resistant colonies are greatly enriched for targeted clones (see *Figure 3*). The degree of enrichment (i.e. the enrichment for targeted colonies achieved by the construct) is determined by comparing the number of neomycin-resistant colonies which arise with the promoterless neo^r construct vs. a similar construct containing a constituitively active enhancer and promoter driving neo^r expression (which should reflect the total frequency of random integrations).

5.2.3 Transcriptionally silent genes

Genes that are not expressed in ES cells can be efficiently targeted by the use of an enrichment scheme termed 'positive–negative selection' (PNS) (63). PNS makes use of the fact that random integration of transfected DNA occurs at its ends, whereas replacement-type homologous recombination seems to occur by a double cross-over event or by gene conversion; in such events, non-homologous DNA located at the ends of the homologous region will be excluded from integration at the targeted locus. Thus, a suitable PNS vector

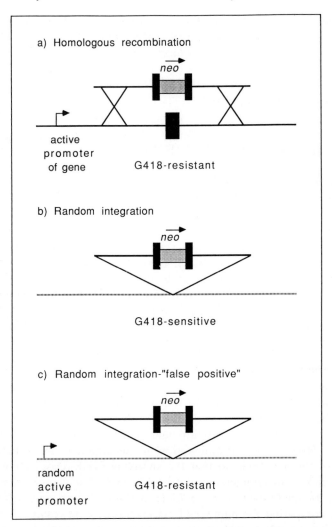

Figure 3. Gene targeting construct for transcriptionally active genes. The neomycin-resistance gene (*neo*) (*hatched box*) without a promoter has been inserted into an exon (*dark box*) within the gene-targeting construct. Homologous recombination between the construct and the endogenous locus will replace the endogenous exon with the promoter-less *neo*-containing exon. Expression of the *neo* gene will be driven by the enhancer and promoter of the endogenous gene leading to a G418-resistant colony (**a**). Most random insertions of the construct will occur into transcriptionally silent regions of DNA and will give rise to G418-sensitive cells (**b**). However, G418-resistant colonies which are not homologous recombinants, 'false positives', will arise after random integration of the construct DNA in close proximity to other active promoters (**c**).

will include the *neo^r* gene, together with an active promoter, inserted into a cloned gene of interest, and an HSV-thymidine kinase gene (*HSV-TK*) located at the end of the genomic clone. All integrants will be resistant to G418 because of the *neo^r* gene. Random integrants will be selectively sensitive to killing by the drug gancyclovir due to the inclusion of *HSV-TK* at the integration site, whereas homologous integrants will not have integrated *HSV-TK* and so will remain resistant to gancyclovir. Double selection with G418 and gancyclovir will thereby enrich for targeted clones (see *Figure 4*). In practice, the efficiency of this technique is quite variable and the enrichment for homologous recombinants is not perfect.

Since PNS requires a *neo^r* gene under independent transcriptional control the efficiency of the positive selection step depends on the efficiency of the *neo^r* expression cassette. One of the most commonly used *neo^r* expression cassettes is pMC1neo (64) which has considerable position dependence in its level of expression (24) causing considerable gene-to-gene variability in targeting efficiency. In addition, this plasmid initially had a mutation which decreased its effectiveness (65), but which has since been corrected in commercially-available lots (Stratagene; lots purchased after about mid 1990). Several other *neo^r* expression cassettes with higher expression than pMC1neo have been reported (for example, refs 24 and 66), which allow an increase in the efficiency of gene targeting.

There is also wide variation in the effectiveness of the negative selection step. The enrichment for targeted cells provided by the *HSV-TK* gene is not absolute, apparently because the gene may be randomly damaged during integration, or its expression may be less than optimal due to the site of integration. In addition, for reasons that are not clear, the reported effectiveness of PNS varies from 10 000-fold to twofold enrichment for targeted ES cell clones in different laboratories. However, there seems to be a consistent achievement of 10–20-fold enrichment in many labs. There has been a report of non-specific toxicity of gancyclovir (66) which would cause an apparent but artificially elevated degree of enrichment. Here, degree of enrichment is determined by comparison of plates selected with G418 alone or with G418 and gancyclovir. For example, an experiment might generate the following numbers: 4×10^7 electroporated cells $\rightarrow$ 2×10^7 survive electroporation (50% survival) $\rightarrow$ 1×10^7 will plate (50% plating efficiency) $\rightarrow$ 1×10^3 will be G418-resistant (1 per 10 000) $\rightarrow$ 50 will be doubly resistant to G418 and gancyclovir (20-fold enrichment) $\rightarrow$ 1 or a few of these will be successfully targeted clones (about 1 per 10^3 integrations).

5.3 Introduction of DNA into ES cells
5.3.1 General considerations
Although there are a number of possible techniques for the introduction of DNA into ES cells for gene targeting, in practice only electroporation is used.

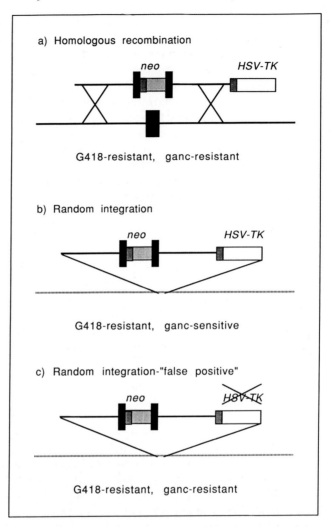

Figure 4. Positive–negative selection gene targeting construct for transcriptionally silent genes. The construct has the *neo* gene (*light-hatched box*), under independent transcriptional control (*dark-hatched box*), inserted into an exon (*dark box*) of the construct as in *Figure 3*. The *HSV-TK* gene (*light box*) is ligated to the end of the homologous region, and is also independently expressed. Replacement-type homologous recombination between the construct and the endogenous gene (**a**) will exclude the *HSV-TK* gene thus giving rise to G418-resistant and gancyclovir (ganc)-resistant colonies. Most random insertion of DNA (**b**) occurs by the ends of the construct and will thereby include the *HSV-TK* gene (as well as the *neo* gene) at the integration site. These cells will be G418-resistant but gancyclovir-sensitive because of the *HSV-TK* gene. Doubly-resistant 'false–positive' colonies (which survive in both G418 and gancyclovir but are random insertions) arise because the *HSV-TK* gene is dysfunctional due to poor expression or damage during transfection (**c**).

Electroporation involves the creation of transient pores in the cell membrane which allow DNA to directly enter the cell. The DNA is relatively undamaged and tends to integrate in the genome as single copies, rather than as concatemers (67). Further, the technique is simple and reproducible and there are several commercially available electroporation units. The chief disadvantages are that, under conditions which maximize the number of integrations, only 50–80% of cells survive and the efficiency of integrations is rather low in that typically only about 1 out of 1000 to 10000 cells will integrate the exogenous DNA (although this value differs considerably between laboratories). A number of protocols have been used for electroporation of ES cells. Below are the techniques which are routinely used in our laboratories and which work well (*Protocol 6*).

5.3.2 Preparation of cells and DNA for electroporation

i. ES cells
The ES cell cultures should be as healthy as possible and rapidly proliferating. Typically, two days before electroporation, ES cell cultures are harvested and counted and plated on to feeder cell monolayers in 75-cm^2 flasks at about $3-5 \times 10^6$ per flask. These should be fed the following day, and fed again 2–3 h prior to harvest for electroporation. Although the total cell harvest is higher after three days of culture, we find that post-electroporation cell survival is considerably reduced.

ii. DNA
The targeting construct plasmid DNA should be as clean as possible but it does not require as stringent purification as is required for nuclear microinjection. Typically, we use plasmid DNA prepared by a standard SDS lysis method since targeting constructs tend to be large (> 10 kb) and require a more gentle preparation technique. The plasmid is purified by caesium chloride gradient centrifugation, but extensive butanol extraction is required to remove all traces of the ethidium bromide. The DNA must be linearized to provide the most efficient substrate for homologous recombination, and the restriction enzyme is inactivated either by heating to 65 °C or by phenol/chloroform extraction of the DNA. To prepare sterile DNA solutions, ethanol precipitate the DNA, wash once with 70% ethanol and dissolve in sterile TE (10 mM Tris, 1 mM EDTA) at 1–2 µg/µl (stored at 4 °C). This solution can be added directly to the cell suspension for electroporation.

5.3.3 Electroporation medium
Several simple types of electroporation media may be used. For example, PBS (pH 7.1) without calcium or magnesium, containing 10 mM Hepes, has worked well for us. Some investigators prefer to use this same solution supplemented with 5 mM KCl, 6 mM dextrose and 0.1 mM β-mercaptoethanol,

whereas others have used DME + 10% FCS and further variations are described in Chapter 2.

5.3.4 Electroporation of ES cells

Protocol 6 is one which we have used with success. However, the various parameters may be varied (for example, by performing all the steps at room temperature) to obtain optimal results in other laboratories. Initially, it is wise to take small aliquots of cells at each step in order to monitor cell viability throughout the procedure, and to help troubleshoot in case problems arise. This simply involves diluting a small volume of cell suspension, plating on to 6-cm Petri dishes, and counting colonies one week later.

Protocol 6. Electroporation of ES cells

1. Harvest the ES cells and wash twice with ice-cold electroporation medium; count cells and resuspend the cell pellet of the last wash to between 1.2 and 2.5×10^7/ml.
2. Dispense 0.75-ml aliquots of the cell suspension into pre-chilled electroporation cuvettes (0.4 cm interelectrode gap) and put on ice. This gives $1-2 \times 10^7$ cells/cuvette.
3. Add 10 to 20 µg of linearized targeting construct DNA to the cold cell suspension and mix well. Leave on ice for 10 min. Just prior to electroporation, mix the suspension well, again taking care not to introduce any bubbles.
4. Electroporate with a single pulse at room temperature at the appropriate settings of capacitance and voltage. (These values must either be empirically determined to maximize the number of neomycin-resistant colonies or based on those previously used with a similar electroporation unit. Typically, these are either 960 or 500 µF and between 200 and 240 V. Recording the time-constant is helpful in that all cuvettes containing similar cell suspensions should have similar discharge characteristics and therefore should have similar values.)
5. The cuvettes are left on ice for a further 10 min, then the contents of each cuvette are suspended in 12 ml of DME + 10% FCS + supplements and are dispersed into three 10-cm Petri dishes already prepared with neomycin-resistance feeder layer and 15 ml of pre-warmed medium.
6. The Petri dishes are then placed in the incubator and fed daily until selection is begun.

5.4 Selection

5.4.1 G418

Selection for G418-resistant colonies is usually begun on the second day after electroporation either by addition of G418-containing medium (DME +

supplements) or by splitting the plate of transfected cells 1:2 or 1:4 on to new plates containing G418. G418 ('Geneticin' Gibco-BRL) working solution is made based on the amount of active drug (expressed as the specific activity: μg active drug/mg actual weight). The necessary dosage to adequately kill susceptible cells in about 7–10 days must be determined empirically, but the drug is usually used between 100 and 400 μg/ml (final concentration) for ES cell selections. G418 can be made as a 100× stock in PBS with 100 mM Hepes (pH 7.4), filter sterilized, aliquoted, and stored at −20°C.

5.4.2 Gancyclovir

Gancyclovir (Syntex; M_r 255) is usually added on the second or third day after electroporation at 2 μM (final concentration). The powder is very stable if kept desiccated. A 1000× stock is made by dissolving 51 mg of the powder in 100 ml sterile H_2O, filter sterilizing, and storing aliquots at 4°C.

5.5 Isolation and characterization of clones

Embryonic stem cells grow as discrete, tightly-packed colonies which adhere to each other much more tightly than to the feeder cell layer. This quality greatly facilitates the isolation of clones. Rather than using cloning rings, entire ES cell colonies (or portions thereof) can be picked and either directly lysed for PCR analysis or dispersed with trypsin in microdrops and added to wells of a 24-well plate. One simply hand-draws a micropipette from a Pasteur pipette, and physically scrapes the desired colony from the surface. While the selection techniques using G418 and gancyclovir enrich for homologous recombinants, final selection must involve individually testing colonies by molecular analysis.

PCR analysis can be performed on portions of colonies by the method described by Kim and Smithies (68) termed the 'boiling-proteinase K digestion-boiling' technique. Primers are chosen which will selectively amplify a recombination-specific fragment (68). Many neo^r gene-specific primers have been reported in the literature in gene-targeting experiments, and much trouble can be saved if one of these is chosen first, and then the primer specific for endogenous gene chosen to match the GC content (and thus the annealing temperature) of the neo^r primer.

Southern blot analysis can be performed during the initial screen of colonies and to verify disruption of the targeted gene in colonies selected as described above. Each well from a 24-well plate can yield between 3 and 10 μg of ES cell genomic DNA for analysis. Restriction enzymes should be chosen to distinguish between the endogenous gene, randomly inserted integrants and successfully targeted homologous recombinant. Ideally, blots should be hybridized with a probe specific for the gene but not found within the targeting construct, then stripped and hybridized with a probe specific for the neomycin coding sequence.

5.6 Production of chimeras

This subject has been extensively covered previously (61). Briefly, aliquots of the targeted ES cell clone are grown as above. Cells are harvested by trypsinization, washed once, and resuspended in DME + 10% FCS + 20 mM Hepes. Blastocysts are harvested by flushing the uterine horns of females at day 3.5 of gestation (where the day of plug is considered day 0.5). Ten to 15 ES cells are microinjected into the blastocoel cavity of each embryo and the injected blastocysts are incubated for one to several hours at 37°C. The blastocysts are then transferred to the uterine horns of day 2.5 pseudopregnant outbred or F1 females, between 6 and 10 embryos to a side. About 17 days later, a litter should be born. The host strain of blastocyst seems to be fairly important. The greatest efficiency for contribution to germline has been obtained by using C57BL/6 host blastocysts, although transmission through germline of ES components using BALB/c and outbred blastocysts has been reported (19, 26). The combination strain 129-derived ES cell → C57BL/6 chimeras are recognizable by the presence of agouti in their coat colour about 10 days after birth. Chimeras derived from BALB/c or other albino blastocysts are recognizable on the day of birth by the presence of darkly pigmented eyes. Once chimeras have been obtained, using selected mutant ES cells, they must be bred to produce a mutant mouse strain. Most likely, the initial mutant will be heterozygous, and may or may not exhibit an obvious phenotype. Breeding must therefore be monitored by appropriate Southern blotting of progeny mice to select carriers of the mutant gene for further breeding and for the production of homozygotes.

Acknowledgements

This work was supported by PHS grants CA29894, CA21124, CA18470, and CA10815 from the National Institutes of Health.

References

1. Martin, G. R. (1980). *Science*, **209**, 768.
2. Solter, D. and Damjanov, I. (1979). *Meth. Cancer Res.*, **18**, 277.
3. Andrews, P. W. (1988). *Biochim. Biophys. Acta*, **948**, 17.
4. Evans, M. J. and Kaufman, M. H. (1981). *Nature*, **292**, 154.
5. Martin, G. R. (1981). *Proc. Natl. Acad. Sci. USA*, **78**, 7634.
6. Robertson, E. J. (ed.) (1987). *Teratocarcinoma and embryonic stem cells: a practical approach*. IRL Press at Oxford University Press, Oxford.
7. Andrews, P. W., Oosterhuis, J. W., and Damjanov, I. (1987). In *Teratocarcinomas and embryonic stem cells: a practical approach* (ed. E. J. Robertson), pp. 207–48. IRL Press at Oxford University Press, Oxford.
8. Stevens, L. C. (1967). *Adv. Morphol.*, **6**, 1.

9. Kleinsmith, L. J. and Pierce, G. B. (1964). *Cancer Res.*, **12**, 154.
10. Mintz, B. and Illmensee, K. (1975). *Proc. Natl. Acad. Sci. USA*, **72**, 3585.
11. Papaioannou, V. E., McBurney, M. W., Gardner, R. L., and Evans, M. J. (1975). *Nature*, **258**, 70.
12. Nicolas, J. F., Dubois, P., Jakob, H., Gaillard, J., and Jacob, F. (1975). *Ann. Microbiol. (Inst. Pasteur)*, **126A**, 3.
13. Martin, G. R. and Evans, M. F. (1975). *Proc. Natl. Acad. Sci. USA*, **72**, 1441.
14. Andrews, P. W. (1984). *Devel. Biol.*, **103**, 285.
15. Pera, M. F., Cooper, S., Mills, J., and Parrington, J. M. (1989). *Differentiation*, **43**, 10.
16. Jones-Villeneuve, E. M. V., McBurney, M. W., Rogers, K. A., and Kalnins, V. I. (1982). *J. Cell Biol.*, **94**, 253.
17. Kubo, Y. (1989). *J. Physiol. (Lond.)*, **409**, 497.
18. Rendt, J., Erulkar, S., and Andrews, P. W. (1989). *Exp. Cell Res.*, **180**, 580.
19. Bradley, A., Evans, M., Kaufman, M. H., and Robertson, E. (1984). *Nature*, **309**, 255.
20. Robertson, E. J. (1987). In *Teratocarcinomas and embryonic stem cells: a practical approach* (ed. E. J. Robertson), pp. 71–112. IRL Press at Oxford University Press, Oxford.
21. Frohman, M. A. and Martin, G. R. (1989). *Cell*, **56**, 145.
22. Capecchi, M. R. (1989). *Science*, **244**, 1288.
23. Robertson, E. J. (1991). *Biol. Reprod.*, **44**, 238.
24. McMahon, A. and Bradley, A. (1990). *Cell*, **62**, 1073.
25. Thomas, K. R. and Capecchi, M. R. (1990). *Nature*, **346**, 847.
26. Joyner, A. L., Herrup, K., Auerbach, B. A., Davis, C. A., and Rossant, J. (1991). *Science*, **251**, 1239.
27. Rossant, J. (1990). *Neuron*, **2**, 323.
28. Damjanov, I., Damjanov, A., and Solter, D. (1987). In *Teratocarcinomas and embryonic stem cells: a practical approach* (ed. E. J. Robertson), pp. 1–18. IRL Press at Oxford University Press, Oxford.
29. McBurney, M. W. and Rogers, B. J. (1982). *Devel. Biol.*, **89**, 503.
30. Edwards, M. K. S. and McBurney, M. W. (1983). *Devel. Biol.*, **98**, 187.
31. McBurney, M. W., Jones-Villeneuve, E. M. V., Edwards, M. K. S., and Anderson, P. J. (1982). *Nature*, **299**, 165.
32. Eddé, B., Jakob, H., and Darmon, M. (1983). *EMBO J.*, **2**, 1473.
33. Pfeiffer. S. E., Jakob, H., Mikoshiba, K., Dubois, P., Guenet, J. L., Nicolas, J. F., et al. (1981). *J. Cell Biol.*, **88**, 57.
34. Fellous, M., Günther, E., Kemler, R., Wiels, J., Berger, R., Guenet, J. L., et al. (1978). *J. Exp. Med.*, **148**, 58.
35. Jakob, H., Boon, T., Gaillard, J., Nicolas, J. F., and Jacob, F. (1973). *Ann. Microbiol. (Inst. Pasteur)*, **124B**, 269.
36. Jetten, A. M., Jetten, M. E. R., Sherman, M. I. (1979). *Exp. Cell Res.*, **124**, 381.
37. Sweeney, T. M., Ogle, R. C., and Little, C. D. (1990). *J. Cell Sci.*, **97**, 23.
38. Darmon, M., Bottenstein, J., and Sato, G. (1981). *Devel. Biol.*, **85**, 463.
39. McBurney, M. W. (1976). *J. Cell Physiol.*, **89**, 441.
40. Muramatsu, T., Gachelin, G., Nicolas, J. F., Condamine, H., Jakob, H., and Jacob, F. (1978). *Proc. Natl. Acad. Sci. USA*, **75**, 2315.

41. Fogh, J. and Trempe, G. (1975). In *Human tumor cells in vitro* (ed. J. Fogh), pp. 115–59. Plenum Press, New York.
42. Andrews, P. W., Damjanov, I., Simon, D., Banting, G., Carlin, C., Dracopoli, N. C., and Fogh, J. (1984). *Lab. Invest.*, **50**, 147.
43. Thompson, S., Stern, P. L., Webb, M., Walsh, F. S., Engstrom, W., Evans, E. P., et al. (1984). *J. Cell Sci.*, **72**, 37.
44. Pease, S., Braghetta, P., Gearing, B., Grail, D., and Williams, R. L. (1990). *Devel. Biol.*, **141**, 344.
45. Martin, G. R., Wiley, L. M., and Damjanov, I. (1977). *Devel. Biol.*, **61**, 230.
46. Lee, V. M.-Y. and Andrews, P. W. (1985). *J. Neurosci.*, **6**, 514.
47. Andrews, P. W., Gönczöl, E., Plotkin, S. A., Dignazio, M., and Oosterhuis, J. W. (1986). *Differentiation*, **31**, 119.
48. Andrews, P. W., Nudelman, E., Hakomori, S.-I., and Fenderson, B. A. (1990). *Differentiation*, **43**, 131.
49. Rathjen, P. D., Toth, S., Willis, A., Heath, J. K., and Smith, A. G. (1990). *Cell*, **62**, 1105.
50. Doetschman, T. C., Eistetter, H., Katz, M., Schmidt, W., and Kemler, R. (1985). *J. Embryol. Exp. Morphol.*, **87**, 27.
51. Robertson, E. J., Bradley, A., Kuehn, M., and Evan, M. (1986). *Nature*, **323**, 445.
52. Hooper, M., Hardy, K., Handyside, A., Hunter, S., and Monk, M. (1987). *Nature*, **326**, 292.
53. Soriano, P., Montgomery, C., Geske, R., and Bradley, A. (1991). *Cell*, **64**, 693.
54. Solter, D. and Knowles, B. B. (1978). *Proc. Natl. Acad. Sci. USA*, **75**, 5565.
55. Shevinsky, L. H., Knowles, B. B., Damjanov, I., and Solter, D. (1982). *Cell*, **30**, 697.
56. Andrews, P. W., Goodfellow, P. N., Shevinsky, L., Bronson, D. L., and Knowles, B. B. (1982). *Int. J. Cancer*, **29**, 523.
57. Kannagi, R., Cochran, N. A., Ishigami, F., Hakomori, S.-I., Andrews, P. W., Knowles, B. B., and Solter, D. (1983). *EMBO J.*, **2**, 2355.
58. Andrews, P. W., Banting, G. S., Damjanov, I., Arnaud, D., and Avner, P. (1984). *Hybridoma*, **3**, 347.
59. Eisenbarth, G. S., Walsh, F. S., and Nirenberg, M. (1979). *Proc. Natl. Acad. Sci. USA*, **76**, 4913.
60. Fenderson, B. A., Andrews, P. W., Nudelman, E., Clausen, H., and Hakomori, S.-I. (1987). *Devel. Biol.*, **122**, 21.
61. Bradley, A. (1987). In *Teratocarcinomas and embryonic stem cells: a practical approach* (ed. E. J. Robertson), pp. 113–151. IRL Press at Oxford University Press, Oxford.
62. Schwartzberg, P. L., Robertson, E. J., and Goff, S. P. (1990). *Proc. Natl. Acad. Sci. USA*, **87**, 3210.
63. Mansour, S. L., Thomas, K. R., and Capecchi, M. R. (1988). *Nature*, **336**, 348.
64. Thomas, K. R. and Capecchi, M. R. (1987). *Cell*, **51**, 503.
65. Yenofsky, R. L., Fine, M., and Pellow, J. W. (1990). *Proc. Natl. Acad. Sci. USA*, **87**, 3435.
66. Tybulewicz, V. L. J., Crawford, C. E., Jackson, P. K., Bronson, R. T., and Mulligan, R. C. (1991). *Cell*, **65**, 1153.
67. Toneguzzo, F., Keating, A., Glynn, S., and McDonald, K. (1988). *Nucleic Acids Res.*, **16**, 5515.
68. Kim, H.-S. and Smithies, O. (1988). *Nucleic Acids Res.*, **16**, 8887.

5

Correlative electrophysiological and biochemical studies in neuronal cell lines

PHILIP M. DUNN, PAUL R. COOTE, and JOHN N. WOOD

1. Introduction

Neuronal excitation is usually evoked by a chemical interaction with a specific membrane-associated receptor that results in an alteration in ion fluxes, a change in the transmembrane potential and a self-propagating wave of depolarization mediated by the activation of voltage-gated ion channels (1). The initial depolarizing event may occur through the direct action of a ligand on a receptor that has intrinsic channel activity, or indirectly through the medium of intracellular second messengers. In addition, the gain of the initial response may be regulated by intracellular second messengers to make the likelihood of action potential propagation more or less likely. The ability to record from single cells, and more recently single channels, has allowed a detailed biophysical description of the properties of some neuronal ion channels to be obtained (2). However, the biochemical events that underly the regulation of activity of voltage-gated and second-messenger-gated channels are less amenable to analysis, because detection methods for alterations in intracellular signalling molecules are fairly insensitive. It is for this reason that clonal populations of cells expressing receptors of interest are particularly useful, as alterations in second-message levels can be unambiguously correlated with alterations in ionic conductances (3).

1.1 Introduction to electrophysiological principles

The electrical properties of a cell result from the dielectric properties of the lipid bilayer of the cell membrane, and the ionic permeability properties of the ion channel proteins inserted within it. Because the intracellular ionic composition usually differs from the extracellular solution, ions will diffuse down their concentration gradient, and a potential difference will develop across the membrane. This potential is dependent on the intra- and extracellular concentrations of ions, and the membrane permeability to the different ionic species, and can be predicted from the Goldman equation:

$$E = \frac{RT}{F} \ln\left(\frac{P_k[K^+]_o + P_{Na}[Na^+]_o + P_{Cl}[Cl^-]_i}{P_k[K^+]_i + P_{Na}[Na^+]_i + P_{Cl}[Cl^-]_o}\right)$$

where P is the permeability coefficient.

A passive cell membrane behaves as a membrane resistance (R_m) and a voltage source (V_m) in parallel with a membrane capacitance (C_m). It is often more convenient to talk in terms of conductance (G) where $G = 1/R$, since conductances in parallel are additive. If a square current pulse is injected into the cell, the membrane potential changes progressively to a plateau level which is given by Ohm's law $\delta V_m = IR_m$. On termination of the current pulse, the voltage decays exponentially with a time-course which is dependent on the values of R_m and C_m:

$$V_t = V_m e^{-t/\tau}$$

where $\tau = R_m C_m$.

If the cell is held under voltage-clamp, and a voltage step is imposed, a large current will initially flow through both R_m and into the membrane capacitance (C_m). Following the transient current, a steady-state current will flow which again will be given by Ohm's law.

Many cells, especially neurons, possess voltage-sensitive ion channels, which will cause a deviation from these simple passive properties, the most spectacular example of which is the generation of action potentials. Furthermore, these descriptions only apply to approximately spherical cells where all the membrane will remain isopotential. The situation in elongated cells, or cells with extensive processes is more complex. For this reason, electrophysiologists find it easier to work with approximately spherical cells with few if any processes, a situation that can be found with many cell lines. A related presentation of electrophysiological analysis of neurons in culture and tissue slices is to be found in *Molecular neurobiology: a practical approach* (4).

Our understanding of the regulation of ion channel activity owes much to advances in molecular genetics. A variety of ligand-gated and voltage-activated ion channels have been identified by expression cloning or by using sequence information derived from conventional protein purification (5, 6). In addition, voltage-activated potassium channels identified in behavioural mutants of Drosophila have allowed a set of homologous neuronal potassium channels to be cloned (7). These studies have revealed a number of structural motifs that are conserved in various ion-channel subsets, and allowed unambiguous studies of the biochemical regulation of these channels to be made in cell lines expressing transfected receptors. Phosphorylation, discussed in detail in Chapter 6 is one important mechanism that is involved in regulation of the activity of a variety of ion channels, and second messengers such as cAMP seem to exert their activity principally through kinase activation. Other mechanisms, for example direct regulatory interactions with channels

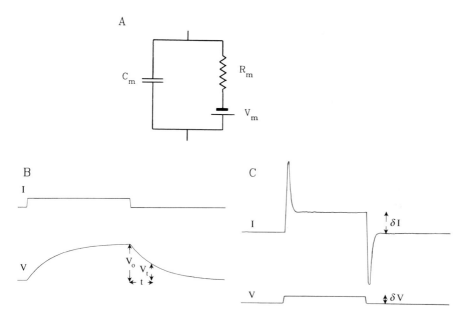

Figure 1. Passive properties of a cell membrane.
A: Circuit diagram for a model cell, consisting of membrane capacitance (C_m) in parallel with membrane resistance (R_m) and potential (V_m).
B: Response of a model cell in current-clamp. Injection of a current pulse produces a change in membrane potential which decays exponentially, so that $v^t = v_0 e^{-t/\tau}$, where $\tau = C_m R_m$.
C: Response of the model cell under voltage-clamp to a voltage command step. There is an initial large transient current (capacity transient) followed by a sustained current where $\delta I = \delta V/R_m$. On termination of the pulse, there is another capacity transient in the opposite direction, before the current returns to the control level.

by cGMP and some eicosanoids, are also known to occur. This chapter is concerned with measuring the best characterized intracellular second messengers, cyclic nucleotides, inositol phospholipids, eicosanoids (arachidonic acid derivatives) and intracellular calcium, and correlating the changes in these molecules with alterations in ionic conductances measured electrophysiologically. *Table 1* contains some examples of second-messenger actions on ion fluxes in a variety of cell types.

2. Modulation of intracellular second-messenger levels

Our present understanding of second-messenger generation has been revolutionized by the discovery that many receptors are coupled to effector systems through GTP-binding proteins (G-proteins) which amplify the signal evoked

Table 1. Examples of actions of intracellular second messengers on ionic conductances in various cell types

Second message	Channel	Mechanism
cGMP	Cation-selective photoreceptor channel	Directly gated (23)
cAMP	Cation-selective olfactory channel	Directly gated (24)
cAMP	Rat heart voltage-gated Ca^{2+} channels	Activated by PKA (25)
$[Ca^{2+}]_i$ increase via inositol phosphates	K^+ channels	Directly gated (26)
Diacylglycerol	Hippocampal K^+ channels	Blocked by PKC (27)
Arachidonic acid metabolites	Aplysia K^+ channel	Directly gated (12-HPETE) (28)
	Rat NMDA receptor	Potentiated (arachidonic acid)

by ligand-binding. The heterotrimeric G-proteins comprise an alpha subunit with GTPase activity that is found associated with beta and gamma subunits in an inactive GDP-binding form. On ligand binding to a receptor associated with a G-protein, the alpha subunit exchanges GTP for GDP dissociates and acts on various targets (enzymes or ion channels) to evoke an effect that terminates when the endogenous GTPase activity of the alpha subunit converts GTP to GDP, and the inactive alpha subunit reassociates with beta and gamma subunits.

The variety of targets acted upon by G-proteins include adenyl cyclase (both activation and inhibition), voltage-gated ion channels, cGMP phosphodiesterases, phospholipase C which produces diacyl glycerol and calcium-mobilizing inositol phospholipids, and phospholipase A2 which can release arachidonic acid and its metabolites. A variety of G-protein subunits have been characterized, and it has become apparent that one subunit may exert multiple activities on, for example, ion channels as well as adenyl cyclase (G_s). In addition, there is redundancy in the system, in that several distinct G-proteins may have the same effects (for example, K^+ channel activation by different G_is (8)).

The involvement of G-proteins in signal transduction can be inferred by the dependence of ligand action on GTP concentrations. Non-hydrolysable GTP analogues such as GTP-γ-S or GppNHp potentiate G-protein-mediated effects. In addition, bacterial toxins which ADP-ribosylate some G-proteins and inhibit their action such as pertussis toxin ($\alpha_i 0$, $\alpha_i 1$, $\alpha_i 2$, and $\alpha_i 3$) or cholera toxin which activates αs can be used to explore the role of particular G-protein subunits in a signalling pathway (9).

The identification of second messengers involved in a signal transduction cascade leading to a conductance change is in some ways more problematic

Table 2. Manipulation of intracellular second messenger levels

(a) Cyclic nucleotides

Dibutyryl cAMP (mM)	diffusible cAMP mimetic
Forskolin (10 μM)	adenyl cyclase activator
Dibutyryl cGMP (mM)	diffusible cGMP mimetic
Nitroprusside (100 nM)	stimulates guanylate cyclase
Rolipram (10 μM)	phosphodiesterase inhibitor
H-7 (10 μM)	general kinase inhibitor
Okadaic acid (1 μM)	phosphatase inhibitor

(b) Inositol phosphates and $[Ca^{2+}]_i$

Neomycin (0.5 mM)	inhibits phospholipase C activity
Ionomycin (0.01–1 μM)	calcium ionophore
A23187 (0.01–1 μM)	calcium ionophore
Caffeine (0.5 mM)	blocks calcium buffering
IBMX (0.1 mM)	blocks calcium buffering
KCl (50 mM) and BayK8864 (1 μM)	activate voltage-sensitive calcium channels

(c) Diacyl glycerol

Phorbol dibutyrate (PDBU-1 μM)	activates protein kinase C
Oleoyl acetyl glycerol (OAG 1 μM)	activates protein kinase C
Phorbol 12, 13 decanoate (PDdec 1 μM)	inactive phorbol control
Neomycin (0.5 mM)	inhibits release of DAG
Staurosporin (100 nM)	blocks protein kinase C

(d) Eicosanoids

Exogenous arachidonic acid (10 μM)	Stimulates eicosanoid metabolism
Exogenous phospholipase A2 (7 units/ml)	mobilizes arachidonic acid
Mepacrine (10 μM)	phospholipase inhibitor
BW755C (30 μM)	lipoxygenase inhibitor
Indomethacin (1–5 μM)	inhibits cycloxygenase pathway
Bromophenacyl bromide (10 μM)	alkylates histidine residues in phospholipase A2
ETYA (arachidonic acid with triple bonds 20 μM)	inhibits eicosanoid metabolism

(e) Nitric oxide

Sodium nitroprusside (100 nM)	production of NO
L-Ng-monomethyl arginine (L-NMMA) 50 μM	inhibits NO generation
L-Ng-nitro arginine (L-NOARG) 50 μM	" " "
L-Ng-nitro arginine methyl ester (L-NAME) 50 mM	" " "
L-Arginine (100 μM)	reverses effect of L-NMMA
D-Arginine (100 μM)	control

than demonstrating the role of G-proteins. The normal method of demonstrating second-messenger involvement is first to demonstrate a ligand-evoked alteration in a particular second-messenger level, and second to try to emulate the effect with reagents that are known to stimulate second-messenger production. *Table 2* lists some reagents known to affect particular second-messenger systems. The failure to identify a second-messenger change

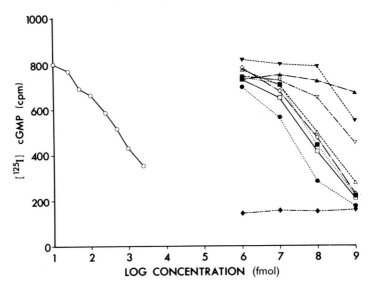

Figure 2. Determination of antibody specificity in a cyclic nucleotide radioimmunoassay. The following nucleotides were testd in a cGMP radioimmunoassay.

Displacement curves of ^{125}I-labelled cGMP from anticyclic GMP serum with various nucleotides

cGMP ○——○	cAMP ●······●	cCMP ▶-----◀
cUMP ▷—··—··—▷	cIMP ♦—··—··—♦	GDP □——□
GMP ■······■	ATP △-----△	ADP ▲—·—·—·—▲

To determine a suitable working dilution for the antiserum, the RIA is performed with 0.7 fmol/tube of [^{125}I]-ligand in the absence of unlabelled ligand using serum dilutions of 1:1000, 1:2500, 1:5000, 1:10 000, 1:25 000, and 1:50 000. Serum dilution is plotted against the percentage bound of the total radioligand added to each tube (% Bo/T) and the antibody concentration that precipitates about 35% of the radioligand under the conditions of the assay is determined. Using this concentration of antiserum, a 'checkerboard' optimization of the assay is carried out by varying the antibody and radioligand concentrations in the range of 1:1.5, 1:2, and 1:3 with or without the addition of 1 fmol of unlabelled ligand. The combination of radioligand and antibody concentrations that shows the highest displacement with cold ligand is then used for the assay. A range of dilutions of unlabelled ligand between 1 fmol and 10 pmol/tube used in the RIA. Even greater sensitivity can be obtained by pre-incubating assays without the radioligand (4–18 h) and then adding the radioligand for a further incubation period (18–24 h).

that correlates with alterations in ion fluxes may mean that a G-protein is directly gating a channel, the ligand is acting through an intracellular mechanism that does not involve soluble messengers (for example, intrinsic receptor tyrosine kinase activation associated with ligand-binding), or that a novel second messenger is at work. Patch-clamp analysis of ligand-induced effects on ion channels can distinguish between these possibilities.

Philip M. Dunn, Paul R. Coote, and John N. Wood

2.1 Measurement of intracellular second messengers

A variety of techniques are used for measuring intracellular second messengers. The cyclic nucleotides cAMP and cGMP are usually measured by radio- or enzyme-linked immunoassay, as it is relatively easy to raise specific high-affinity antisera to these compounds. Convenient commercial kits for RIA or EIA are available from numerous commercial sources, including Amersham, New England Nuclear, and Cayman Chemicals. Such kits can become a major expense if used extensively, and we therefore include a protocol for raising highly specific antisera to cyclic nucleotides (*Protocol 1*) which can be used to develop cheap in-house immunoassays.

Raising specific antisera to lipid mediators such as eicosanoids or diacylglycerol is more problematic, due first to the difficulties in coupling compounds to carrier molecules, and second because of the range of structurally related compounds that may cross-react with the antiserum. Although many specific radioimmunoassays have been developed for particular eicosanoids, there is still no such assay for arachidonic acid itself. For this reason, these compounds are generally detected by chromatographic analysis after radio-labelling with an appropriate precursor if the whole range of arachidonic acid metabolites is to be analysed. Inositol phosphates can be measured either by immunoassay or through ion-exchange chromatographic analysis or HPLC after cell labelling with myo-[^{3}H]-inositol. DAG release can be measured directly by TLC using ^{14}C-arachidonic acid to label phospholipid precursors (10), although stimulation of protein kinase C activity is a more frequently used indirect method of measuring DAG release. TLC or HPLC are also frequently used for measuring eicosanoids (11). Nitric oxide, the most recently discovered second-messenger molecule, is still only detectable indirectly in bioassays.

Protocol 1. Raising and testing highly specific antisera for cyclic nucleotides

Raising antibodies and carrying out test bleeds in rabbits must only be carried out by fully trained personnel in accord with local legal requirements.

1. Conjugate 20 mg of O-2'-monosuccinyl guanosine 3'5'-cyclic monophosphoric acid (sodium salt, Sigma), (or O-2'-monosuccinyl adenosine 3'5'-cyclic monophosphoric acid) to 230 mg thyroglobulin (Sigma) with 12.2 mg of the water soluble carbodiimide 1-ethyl-3-dimethylamino-propyl-carbodiimide (EDAC, Pierce) in 5 ml 10 mM Tris–HCl pH 7.2 overnight at room temperature.

2. Dialyse the conjugates extensively (three changes minimum) against large volumes of 10 mM Tris–HCl pH 7.2 at 4°C and store aliquots of the conjugate at −70°C.

Protocol 1. *Continued*

3. Inject rabbits intramuscularly with 1 mg of conjugate emulsified in Freund's complete adjuvant. Other adjuvants (for example, Poly rI. poly rC) may be substituted.
4. Repeat the immunization at monthly intervals with incomplete adjuvant, and 5 days after the fourth immunization, take a test bleed for analysis of specificity in a competition radioimmunoassay using various nucleotide phosphates (see *Protocol 3*) and antibody concentrations as described in the legend to *Figure 2*.

Protocol 2. Sample preparation for measurement of cyclic nucleotide levels by immunoassay

1. Soak 13 mm glass coverslips with 100 µg/ml poly-D-lysine (M_r 100 000) overnight, wash them twice in sterile water and dry them in a laminar flow hood. UV-irradiate them, and store in a sterile 24-well tissue culture plate.
2. Plate 10 000–20 000 dissociated cells on to the coverslips in normal growth medium and leave them to grow overnight or differentiate according to selected culture media.
3. Lift individual coverslips, using forceps and a microspatula, and briefly suspend them in assay buffer at 37°C. Assay buffer is Hepes-buffered Hanks' solution or equivalent buffered isotonic saline pH 7.4 at 37°C.
4. Suspend the coverslip in assay buffer containing the ligand under investigation at an expected supramaximal dose for specified times (10, 20, 30, 40, 60, 90, 120, 300 sec).
5. Immerse the coverslips in 1 ml acid alcohol (99% ethanol/1% 1 M HCl) in a 24-well plate to stop the reaction.
6. After a minimum time of 2 h remove the supernatant and dry down under nitrogen at 60°C prior to radioimmunoassay.
7. Treat the coverslips overnight with 1 M NaOH, neutralize the solution and assay protein by the Bradford method (see *Protocol 15*, Chapter 8).
8. Resuspend the samples in buffer for enzyme-linked or radioimmunoassay (*Protocol 3*).
9. Using the time point that gave a maximal response, vary the concentration of agonist to determine the EC_{50} value for cyclic nucleotide induction.

It is wise to carry out each experimental point in triplicate. Do not plan experiments using too many samples. Cyclic nucleotide samples, either lyophilized or in acid-alcohol are stable for days.

Protocol 3. Measuring cAMP and cGMP by radioimmunoassay

1. Prepare standard solutions of cyclic nucleotides (*Figure 2*) in assay buffer which is 50 mM sodium acetate pH 6.2.
2. Incubate 100 μl volumes of standard solutions with 100 μl antibody (1:20 000 dilution) and 100 μl iodinated cAMP or cGMP tracer (NEN, 2200 Ci/mmol, 15 nCi/ml in sodium acetate pH 6.2 containing 0.15% BSA) at 4°C overnight in plastic test-tubes (LP3, Luckham).
3. Add 100 μl of Sac-Cel anti-rabbit cellulose (IDS) suspension to each tube to precipitate the immune complex of antibody and ligand, vortex briefly, and incubate the samples at 4°C for a further 30 min.
4. Centrifuge the samples at 1500 g for 30 min at 4°C.
5. Aspirate the supernatants, leaving the precipitated radiolabelled tracer in the packed Sac-Cel pellets.
6. Count the pellets in a gamma-counter equipped with a radioimmunoassay protocol.

Protocol 4. Measuring eicosanoid and diacylglycerol release

1. Prepare coverslips of cells as described in *Protocol 1*.
2. Label the cells for a minimum time of 2 h with [^{3}H]-arachidonic acid (Amersham 80 Ci/mmol) in growth medium (20 μCi/ml). (The complexities of eicosanoid metabolism make it very difficult to reach conditions of isotopic equilibrium).
3. Wash the coverslips of cells four times with Hepes-buffered Hanks' solution pH 7.4 at 37°C.
4. Place the coverslips in the same medium containing the drug of interest for a specific time (for example, 30 nM bradykinin, 5 min).
5. After incubation at 37°C, immerse the coverslips in glass tubes containing 1 ml chloroform:methanol 1:1 v/v for 15 min and then add 1 ml of water.
6. Vortex the tubes and extract the organic phase after brief centrifugation.
7. Dry the organic phase under nitrogen and resuspend the dried material in 1–200 μl chloroform and analyse it by thin-layer chromatography (TLC) on Whatman LK5D plates. Spot 50 μl samples at the base of the TLC plate.
8. (a) To detect arachidonic acid metabolites use a solvent system of ethyl acetate: 2,2,4-trimethylpentane:acetic acid:water (110:50:20:100 by volume). Use [^{14}C]-arachidonic acid and other appropriate markers (15).

Correlative electrophysiological studies in neuronal cell lines

Protocol 4. *Continued*

    (b) To detect diacylglycerol labelled with [^{14}C]-arachidonic acid use a solvent system of benzene:diethyl ether:ethyl acetate:acetic acid (80:10:10:0.2 by volume) plus 0.01% butylated hydroxytoluene (16). Use 1-stearyl-2-[^{14}C]-arachidonyl glycerol and [^{14}C]-arachidonic acid as markers.

9. When the solvent front approaches the top of the TLC plate, remove it and dry the plate in a fume-hood.
10. Radioautograph the TLC plate using Hyperfilm β-max at $-70°C$ for 2–10 days. Identify radiolabelled bands after film development by comparison with standards and known relative migration values (15).
11. Elute the identified bands from the TLC plate by scraping the silica into a scintillation vial and adding 1 ml diethyl ether and scintillant. Count the samples in a scintillation counter.

Protocol 5. Measuring inositol phospholipid turnover (17)

1. Prepare coverslips of cells according to *Protocol 2*.
2. Radiolabel the cells overnight in growth medium containing myo-2-[^{3}H]-inositol (50 µCi/ml, Amersham ICN or NEN).
3. Wash the coverslips of cells in Hepes-buffered DMEM (37°C) as described in step 4 of *Protocol 2*.
4. Suspend coverslips in medium containing the ligand of interest, for 0, 10, 20, 40, 60, and 180 sec. Carry out each time-point in triplicate.
5. Immerse the coverslips in 1 ml of ice-cold 10% trichloracetic acid to stop the reaction.
6. After 20 min, extract the cells with diethyl ether to remove unincorporated label, add 50 µl of phenol red to each sample as a pH indicator and neutralize the samples with 1 M ammonium hydroxide.
7. Separate the labelled inositol phosphates by ion-exchange chromatography on 1.5-ml columns of anion exchange resin Dowex-1 in its formate form (200–400 mesh), prepared by sequential 10-column volume washes with 1 M NaOH, water, 1 M formic acid, and water checking that the pH is greater than 4.5 before applying samples.
8. Apply the sample in water, and collect 5-ml batches of eluted material directly into large volume (20-ml) scintillation vials.
9. Sequentially elute, using 2 × 5 ml volumes:
    (a) the free inositol with 5 mM inositol
    (b) glycero-inositol phosphate with 5 mM sodium tetraborate/60 mM ammonium formate

(c) inositol phosphate with 200 mM ammonium formate in 0.1 M formic acid
(d) inositol diphosphate with 450 mM ammonium formate in 0.1 M formic acid
(e) inositol trisphosphate/0.1 M formic acid with 0.75 M ammonium formate in 0.1 M formic acid
(f) inositol tetrakisphosphate with 1 M ammonium formate in 0.1 M formic acid
(g) Add water soluble scintillant (for example, Amersham ACS) to fill the vials and count the samples using a quench corrected d.p.m. programme in a liquid scintillation counter.

10. Compare the profile of counts released in the various fractions between stimulated and control cultures. It may be convenient to present the data as % of counts present in phosphatidyl inositol.

2.2 Biochemical measurement of ion fluxes

A variety of optical probes (Chapter 7) are now available that enable free intracellular calcium levels to be calculated (Molecular Probes Inc.). A simple index of intracellular calcium mobilization that does not require complex equipment is the release of ^{45}Ca from cell lines that may reflect agonist-induced elevation in calcium levels mobilized from intracellular stores, or increased permeability to calcium. The same protocols can be adapted to measure ^{86}Rb efflux which is permeant in a variety of potassium channels and thus provides a measure of potassium flux. ^{42}K has a half-life of 12 h and is thus almost impossible to work with, whilst ^{86}Rb has a $T_{1/2}$ of 19 days. Sodium flux can be measured directly with the beta and gamma emitter ^{22}Na ($T_{1/2}$ = 2.6 years). However, ^{14}C-guanidine (Amersham) is a lower-energy, longer-lived compound which is also permeant through some sodium channels, and which therefore provides a convenient marker of sodium fluxes. The ion flux assays described here are useful measures of the biochemical actions of various agents on a population of cells. They are thus a complement to single-cell electrophysiology and can give quantitative information simply and rapidly, particularly with respect to determining the dose-dependency of a ligand-evoked alteration in ion fluxes (18). In addition, external manipulation of second messenger levels with various agents (*Table 1*) can give useful information on those intracellular agents involved in regulating a particular ligand-evoked conductance change. *Protocol 6* details the procedure for measuring calcium efflux, but the same method can be used with ^{86}Rb or ^{14}C guanidine. In *Figure 3* we can see that biochemical measurements of ion fluxes in a bradykinin-sensitive sensory-neuron-derived cell line can provide considerable information about the mechanism of action and the attendant ion fluxes

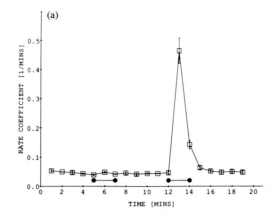
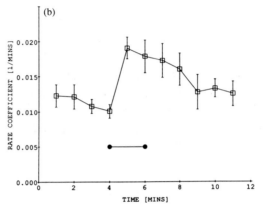
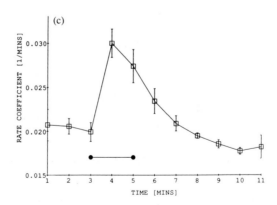

Figure 3. Bradykinin-evoked ion fluxes in a neuronal cell-line. (**a**) ^{45}Ca efflux evoked by 0.1 μM PDBU (5–7 min) and 0.3 μM bradykinin (12–14 min) from differentiated ND7/23 cells. (**b**) ^{86}Rb efflux from DRG neurons evoked by 0.3 μM bradykinin (4–6 min), and (**c**) ^{86}Rb efflux evoked by bradykinin (3–5 min) in sensory-neuron-derived ND7/23 cells.

evoked by bradykinin on both the model cell line and the neurons from which the line was constructed.

Application of a phorbol ester to the cell line (*Figure 3a*) prior to addition of bradykinin neither mimics nor desensitizes the bradykinin-evoked calcium efflux, showing that protein kinase C is not involved in this response or its desensitization. Bradykinin, however, does show desensitizing calcium-mobilizing responses in this line, through a mechanism that therefore must not involve protein kinase C. Both cell line and sensory neurons (*Figure 3b* and *c*) show a ^{86}Rb efflux evoked by bradykinin, that reflects increased potassium efflux, probably through the activation of calcium sensitive potassium channels. Note how much more robust the response in *Figure 3c* is using the cell line, when compared with the smaller effect observed in the parental sensory neurons (*Figure 3b*), that are extremely difficult to obtain in adequate numbers for such experiments.

Protocol 6. Measuring ^{45}Ca efflux from cells in culture

1. Plate the cells into 60 × 10 μl wells of a Terasaki plate that has been pre-coated with poly-D-lysine in the same way as coverslips in *Protocol 1*.
2. Load the cells with isotope (^{45}Ca 10 μCi/ml or equivalent concentrations of ^{86}Rb or ^{14}C guanidine) in growth medium at 37°C for 2 h.
3. Wash the cells rapidly with 3 × 10 ml assay buffer pipetted against the side of the Terasaki plate, gently rocking the plate, and discarding the buffer into a beaker.
4. Place the plate on a dry block set at 37°C and change the medium each minute for 7 min, decanting 10 ml of buffer into a scintillation vial.
5. Add assay buffer containing agonist for two consecutive time-points.
6. Continue washing and collecting the buffer at 1-min intervals for another 7 min.
7. Finally, add 10 ml 0.1% SDS to the plate for 10 min and count the total number of counts left in the cells in aqueous scintillant (for example, Amersham APS). Express the counts released as a rate constant by dividing the number of counts released in any time-period with the total number of counts left in the cells at that time point.

---

3. Electrophysiological effects of neuroactive ligands

3.1 Electrophysiological actions of neurotransmitters, neuromodulators, and drugs

The majority of electrophysiological effects of neuroactive ligands, referred to here as drugs, result from an increase or decrease in the membrane permeability to particular ions, resulting in a depolarization or hyperpolarization. If the depolarization is sufficient to evoke action potential generation, calcium influx at the synapse results in the release of neurotransmitter and the transmission of a chemical signal to the postsynaptic site of a neighbouring cell. Transmitters that directly gate ion channels show rapid effects on membrane permeability (milliseconds) whilst transmitters that act indirectly through second messengers may take minutes to act (14). The most thoroughly studied example of fast transmitter action is provided by studies of the effect of agonists on the nicotinic acetylcholine channel/receptor complex (15).

Activation of the receptor opens an ion channel which is equally permeable to K^+ and Na^+ ions. A cell which is equally permeable to Na^+ and K^+ ions will have a membrane potential close to 0 mV, so that activation of the nicotinic receptor causes the membrane potential to shift towards this reversal potential—a term used to denote the voltage at which the current flow across the cell membrane changes direction. Activation of the inhibitory chloride channel gated by GABA, however, allows an increase in Cl^- flux across the cell membrane which leads to a shift towards the chloride equilibrium potential. This may, in fact, be positive or negative to the membrane potential (depending on the cell type), so depolarization or hyperpolarization may occur, but the decrease in membrane resistance reduces the excitability of the cell.

Apart from fast transmitters and drugs acting directly on receptor channel complexes, there are many examples of indirect drug actions on ion channels, often mediated through G-protein-induced alterations in second messengers. The neuron may thus receive a variety of external signals which determine its responsiveness. Slow acting drugs may also activate as well as modulate cation or anion selective channels. Of particular significance to neuronal function are those drugs that modulate the voltage-sensitivity of calcium channels or the activity of the excitatory amino-acid-gated NMDA channel. Either action can profoundly alter the functional consequences of neuronal excitation by modulating the strength of synaptic coupling.

3.2 Electrophysiological characterization of drug responses

In order to compare the response evoked in a cell-line with that observed in primary cultures, or with responses evoked by putative second messengers, it is necessary to characterize the response in as much detail as possible, either under current-clamp, or (preferably) under voltage-clamp. The properties of the response which should be examined are listed below.

- Does the drug evoke a hyperpolarization or depolarization (an outward, or inward current under voltage-clamp)?
- Is there an increase or decrease in membrane resistance?
- What is the reversal potential for the response?

- What is the ionic dependence of the response, i.e. how does ion substitution with impermeant ions affect the response amplitude and reversal potential?
- Is there any voltage-sensitivity of the response (i.e. is the current voltage-relationship for the response non-linear)?
- How is the response affected by known ion-channel blockers?

4. Electrophysiological recording techniques

The development of the glass microelectrode by Ling & Gerard in 1949 (16) has enabled physiologists and pharmacologists to investigate the electrical properties of a wide variety of cell types. There are now several different recording techniques which can be used to study changes in the electrical parameters of the whole cell, or changes in the properties of single ion channels present in patches of cell membrane (patch-clamp). The techniques used can be divided into two types

(a) 'sharp microelectrode' recording
- voltage recording
- voltage and current passing (current clamp)
- voltage-clamp

(b) 'patch-clamp' recording
- whole-cell voltage-clamp
- cell-attached patch (single-channel recording)
- outside out patch (single-channel recording)
- inside out patch (single-channel recording)

Each technique has advantages and disadvantages. Voltage-clamp recording provides more information than current clamp recording but can be more difficult to carry out. Sharp microelectrodes can be used for recording in intact tissues or dissociated/cultured cells. The electrode filling solution produces minimal changes in the composition of the cytosol. Patch-clamp recording requires a clean membrane surface, and is usually carried out on dissociated/cultured cells (more recently, patch-clamp recording has been carried out in thin brain slices; ref. 17). With whole-cell patch-clamp recording, the cytosol is dialysed with the electrode filling solution, thus the ionic composition of the cytosol can be modified, and membrane impermeant putative second messenger candidates can be introduced into the cell. While this is advantageous, dialysis may also result in the loss of cytosolic components vital for the response under consideration. The isolated patch techniques provide ready access to either the cytoplasmic or the extracellular face of the membrane, to which putative second messengers, or agonists can be applied respectively, although again there is the possibility of the loss of vital cytoplasmic components.

4.1 Equipment and practical considerations

(a) Electrode pullers

Glass microelectrodes are made by heating capillary tube until it softens, when it is rapidly stretched as it cools down. The tip draws out, breaks, and separates. There are a number of designs of electrode puller manufactured for this process. The heat is produced by passing electric current through a metal filament. The pull is produced either by gravity, or by an electromagnet.

- Vertical pullers use gravity or electromagnets to produce the pull. These usually produce quite coarse (low resistance <40 MΩ) electrodes, and are often used in two stages to produce electrodes for patch-clamp recording (for example, List Medical or Sutter Instrument Co.).
- Horizontal pullers of the Livingstone type use a spring and gear mechanism. They are good for fine microelectrodes 30–300 MΩ, but produce rather a long and whispy shank (for example WPI).
- Horizontal pullers of the Narishige type use an electromagnet to generate the pull. They also produce fine electrodes with a long shank (Narishige, or Campden Inst.).
- Horizontal pullers of the Brown and Flaming type are electromagnet-driven, with a gas jet which rapidly cools the heating filament. They are good for fine electrodes with short shank, but setting them up correctly can be very time-consuming (Sutter Instrument Co.).

For sharp microelectrode recording, the size and geometry of the electrode tip will greatly influence the amount of damage done to the cell on penetration. However, there are physical and electronic constraints on the tip diameter. In particular, the smaller the tip diameter, the greater the electrical resistance. Electrodes typically have resistances in the range 20–200 MΩ. Resistors of this magnitude make good aerials, and the signal recorded by the electrode will be contaminated by radio signals. (Interference problems of this type can be reduced by mounting the recording set up in a metal cage electrically connected to earth.) Patch electrodes are usually pulled in two stages on a vertical puller, to produce a steep taper to a tip diameter of 1–5 μm. To reduce the capacitance to ground, the tip of the electrode is often coated with Sylgard silicone resin (This is carried out prior to fire-polishing.)

(b) Microforge

Once pulled, micropipettes for patch-clamp recording are usually fire-polished, to produce a smooth and very clean tip. This is done by bringing the tip of the electrode to within a few microns of a heated platinum-iridium wire filament under visual control at $>\times 200$ magnification. Commercially produced microforges are available for this task (Narishige, or List Medical).

(c) Microscope

To record from a single cell in tissue culture requires an optical system with sufficient magnification and resolution to be able to view the cell, and the tip of the microelectrode. For most applications, an inverted microscope is the instrument of choice (for example, Nikon TMS or Diaphot).

(d) Air table

Movement of the microelectrode with respect to the cell may produce excessive damage to the cell, or loss of the impalement. For these reasons, sources of mechanical disturbance and vibration must be minimized. This is usually achieved by mounting the recording set-up on some form of air-cushioned anti-vibration table.

(e) Micromanipulator

For precise positioning of the electrode tip onto a cell, some sort of micromanipulator is required, either mechanical (Leitz), or hydraulic (for example, Narishige).

(f) Amplifiers

For sharp microelectrode recording, the amplifier is based on a voltage-follower. It should have an input resistance 100–1000 times that of the recording microelectrode, and a leakage current less than that required to produce a 1 mV change in the membrane potential. In addition, the response time of the amplifier should be more than 5 times faster than the signal of interest. Numerous commercial models are available (for example, Neurolog NL107). For patch-clamp recording, the amplifier is based on a current to voltage converter. Most commercially available amplifiers have switchable feedback resistors which enable them to be used at low sensitivity (1 mV/pA) for whole-cell recording, or high sensitivity (100–1000 mV/pA) for single-channel recording (List EPC7, Axon Inst. Axopatch).

(g) Ancillary equipment

In addition to the above items of equipment, an oscilloscope, a chart recorder, and some type of pulse generator are also required for a basic electrophysiology.

4.2 Sharp microelectrode recording

For these techniques, the recording microelectrode has a tip diameter of $< 1\ \mu m$, and is back-filled with an electrolyte solution (usually 1–3 M KCl) using a syringe and fine needle. Filling to the tip occurs by capillarity, and is facilitated by the inclusion of a glass fibre in the lumen of the capillary tubing during manufacture.

4.2.1 Current-clamp recording

While monitoring membrane potential will give some indication of a drug's action, the amount of information generated is very limited. Further information can be obtained if the membrane potential can be displaced by injecting current into the cell. By injecting current pulses of a known amplitude (δI) and monitoring the change in membrane potential (δV), the resistance of the cell membrane (i.e. the input resistance (R_m)) can be determined (since by Ohm's law $R_m = \delta V/\delta I$). This will then indicate whether the drug's action is due to the opening or closing of membrane ion channels. Current injection may be carried out using a second microelectrode connected to a suitable current source, or via the recording electrode. In the latter case, additional electronic circuitry, in the form of a 'bridge circuit' is used to subtract from the voltage signal the potential change produced by passing the current through the high resistance of the microelectrode.

Most commercially available amplifiers for intracellular recording incorporate a 'bridge circuit' to permit current injection through the recording electrode to allow so-called current-clamp recording to be carried out.

The impalement of cells is described in *Protocol 7*.

Protocol 7. Impaling cells with microelectrodes

1. Fill the electrode with electrolye solution, insert into electrode holder (electrical contact is established between the filling solution and an Ag/AgCl pellet in the electrode holder).
2. Lower the electrode tip into the bath, and manipulate the electrode towards the cell.
3. Adjust the amplifier offset, to give an output of 0 mV, to null any potential difference between the recording electrode and the bath electrode (due to electrode potential and tip potential).
4. While passing small current pulses (0.1 nA, 50 ms, 10 Hz), adjust the bridge balance, so that the current pulse produces no deflection.
5. Adjust the negative capacitance to remove as much of the capacity transients at the beginning and end of the current pulse as possible, without the amplifier oscillating.
6. Advance the microelectrode until it touches the cell. At this point, the electrode resistance will increase, and the current pulse will evoke a change in the voltage trace.
7. Impalement is best accomplished by oscillating the amplifier by overcompensating with the negative capacitance very briefly. Alternatively, tapping the micromanipulator, or advancing the electrode rapidly 1–2 μm by means of a stepper motor, will impale the cell.

4.2.2 Voltage-clamp

Recording membrane potential is useful, but has severe limitations. The change in membrane potential is not linearly related to the number of ion channels activated, but is limited by the reversal potential of the response. The time-course of the response is limited by the membrane time-constant, which may cause distortion of fast responses. Furthermore, change in membrane potential by the drug action may lead to the activation (or inhibition) of other voltage-sensitive membrane ion channels, thus complicating the response. One way around these problems is to hold the membrane at some predetermined potential and monitor the transmembrane current required to achieve this. This technique is known as voltage-clamp recording. The essential electronics for a voltage-clamp system are a voltage monitor, to measure the membrane potential; a feedback amplifier which compares the measured potential (V_c) with the command potential (V), and outputs a signal proportional to the difference, which is then used to control the output of the current injection circuit. The first voltage-clamp circuits used separate potential measuring and current injecting electrodes. However, there are now amplifiers available which enable voltage-clamp to be carried out using a single microelectrode (for example, Axoclamp, Axon Inst.). Most of these work by switching at high frequency between current-passing and voltage-recording modes.

By determining the current flowing at different membrane potentials (either by applying voltage steps, or a slow voltage ramp) current–voltage relationships (CVRs) can be obtained. By subtracting the CVR obtained in the absence of drug from one obtained in the presence of drug, the CVR for the drug-activated current can be determined, and the reversal potential (potential at which zero current flows) estimated, giving further information about the ionic selectivity of the drug-activated current.

4.3 Determination of current–voltage relationships and reversal potential

While observing a response and monitoring the conductance change can be informative, further electrophysiological characterization of the response requires determination of the reversal potential and the current voltage relationship of the response. Although these parameters can be determined under current clamp, the experiments are best performed under voltage clamp. There are essentially two ways to carry out the experiment, depending on the duration and reproducibility of the response being investigated.

For reproducible, short-duration responses:

(a) Record responses while holding the membrane potential at a number of different potentials.

(b) Plot a graph of response amplitude vs. membrane potential.

(c) Read off the membrane potential at which there is no response. It may not always be possible to clamp the cell at potentials positive and negative to the reversal potential, in which case the reversal potential must be estimated by extrapolation.

For well-maintained responses:

(a) Record current responses to a series of voltage command steps (i.e. a control current voltage relationship).
(b) Apply the drug, and repeat the same voltage command steps.
(c) Plot the current–voltage relationships obtained in the presence and absence of the drug. The intersection of the two will yield the reversal potential for the response.
(d) Subtraction of the control current levels from those in the presence of the drug will yield the CVR for the drug response, which will cross the zero current axis at the reversal potential.

The use of computers to produce voltage commands has made it easy to produce linear voltage ramp commands, which provides a more rapid means of obtaining a current voltage relationship.

4.4 'Patch-clamp' recording

This technique was first described in 1976 by Neher and Sakmann, and subsequently modified and improved (18). In these techniques, a glass pipette with a fire-polished tip of 2–5 μm in diameter is pressed against the cell membrane, so that a tight seal with an electrical resistance of > 1 GΩ (i.e. a gigaseal) is formed between the glass and the membrane. The electrode is filled with a physiological salt solution, and connected to a very sensitive current–voltage converting amplifier. This permits the potential of the electrode to be fixed, and the flow of ionic current through single ion channels can be monitored. This procedure is described in *Protocol 8*.

Protocol 8. Obtaining a gigaseal for patch-clamp recording

1. Fill the electrode with the appropriate physiological salt solution (for cell attached and inside out recording this may be normal Ringer's solution, while for whole cell or outside out a solution of the composition given below may be used), and fit it into the electrode holder. (For patch-clamp recording, the electrode holder has a side port for application of suction, and electrical contact to the electrode filling solution is made via a chlorided silver wire.)

2. Lower the electrode into the bath, and set the current to zero by adjusting the voltage offset control (to correct for electrode potentials).

3. Apply 10 mV voltage command steps (20 msec 10 Hz) and bring the electrode tip into contact with the cell; this is indicated by a rise in a pipette resistance (i.e. a decrease in the current excursion to the voltage step).
4. Apply gentle suction (10–50 cm of water) to the pipette. The pipette resistance should continue to rise to $> 1\,G\Omega$. (i.e. $< 10\,pA$ current excursion). This may happen immediately, or may take some time. Seal formation may be facilitated by applying a steady negative holding potential of 40 to 60 mV. Inclusion of an appropriate concentration of agonist in the pipette solution will enable activation of ligand-gated ion channels in the membrane patch.

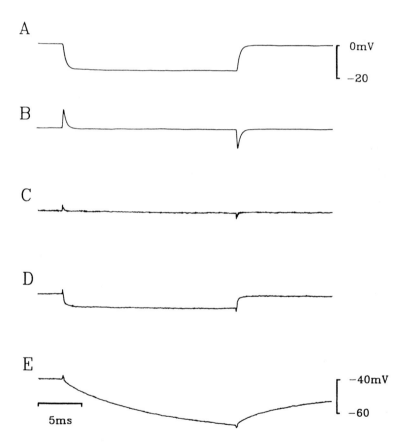

Figure 4. Impalement of a cell with a microelectrode. The traces are voltage records of the response to a 0.2 nA current pulse, at different stages of the impalement procedure. **A:** electrode in the bath with no negative capacitance and no bridge balance compensation. **B:** after adjustment of the bridge balance. **C:** after subsequent adjustment of the negative capacitance. **D:** touching the cell. **E:** successful impalement.

In addition to recording current flow through single ion-channels in this cell-attached configuration, a number of other configurations can also be established.

(a) By applying suction to the recording pipette, the membrane patch can be ruptured, giving ionic and electrical access to the interior of the cell. In this 'whole cell' voltage-clamp mode, all the cell membrane is voltage-clamped, and total membrane current can be monitored.

(b) By withdrawing the electrode from the cell after establishing the cell-attached configuration, an isolated patch of membrane can be obtained, with the cytoplasmic face exposed to the bathing solution. Frequently, a small membrane vesicle forms at the end of the electrode which has to be ruptured by exposure to the air to uncover the cytoplasmic face. This is the inside-out configuration.

(c) By withdrawing the recording electrode after establishing the whole-cell configuration, an outside-out membrane patch is formed.

Typical pipette-filling solution for whole-cell recording is:

- KCl   140 mM
- Hepes   10 mM
- EGTA   10 mM
- CaCl   1 mM

pH to 7.2 with KOH

4.5 Methods of drug application

There are several techniques used to apply drugs to cells during electrophysiological recording, each has its advantages and disadvantages. Whichever method is employed, it is necessary to ensure that it does not introduce mechanical or electrical disturbance to the recording set-up.

While bath application is simple, the responses are usually relatively slow in onset and offset. Iontophoresis (*Protocol 9*) can produce very rapid

---

Figure 5. Determination of the reversal potential of a response.
A: for the response of voltage-clamped ND7/23 cell to 5-HT. (*i*) shows responses to 10 μM 5 HT recorded at membrane potentials of −80, −60, and −10 mV. (*ii*) shows a graph of response amplitude vs. membrane potential for the responses, which gives a reversal potential of +2 mV.
B: for the response of an ND7/23 cell to bradykinin. (*i*) Current responses to a series of voltage command steps recorded in the absence of bradykinin. (*ii*) A graph showing current–voltage relationships obtained by plotting the absolute, steady-state current as a function of the membrane potential during the voltage step, from the data in (*i*) (21), and from a second series of pulses recorded in the presence of bradykinin. The intersection of the two lines yields a reversal potential of −20 mV.

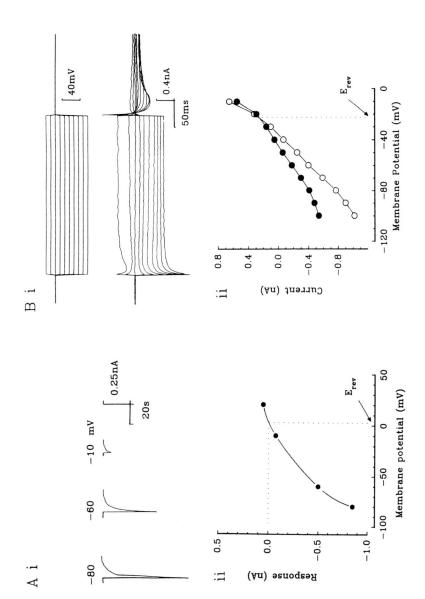

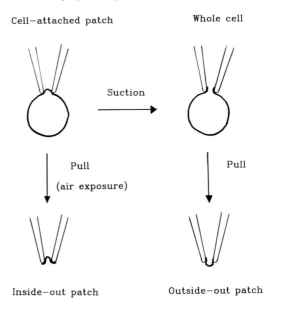

Figure 6. Formation of the four different patch-clamp recording configurations. The patch electrode is initially sealed on to the membrane to form the cell-attached configuration. Application of suction gives rise to whole-cell recording. Subsequent withdrawal of the electrode produces an outside-out isolated patch. Withdrawal of the electrode from the cell-attached configuration produces an inside-out patch.

responses, but only works for charged ligands, and the drug concentration is not known (although it can be estimated). Puffer application can give rapid responses, and the drug concentration is known. However, it is usually possible to apply only one or two drug solutions. Microperfusion techniques, while more complex to set up, combine fairly rapid application of known drug concentrations with the ability to apply a number of different drug solutions.

4.5.1 Bath application

This is the simplest technique. Simply change the inflow to the bath so that it is perfused with a solution containing drug at the required concentration. The change over can be carried out using manual, or electrically operated valves.

Protocol 9. Iontophoretic drug application to cells

1. Fill a glass microelectrode with a concentrated solution of drug in an ionized form (this may require pH adjustment of the solution).
2. Mount the electrode in a holder, connected to a suitable current-passing source.
3. Position the tip of the electrode close to the cell of interest, while passing a

small holding current through the electrode to prevent unwanted drug leakage (negative current for a positively charged drug).

4. Apply current pulse to eject drug (positive current for positively charged drug). The amount of drug ejected is proportional to the current passed, and the duration of the current pulse. The use of multi-barrel electrodes permits several drugs to be applied to the same cell. This technique can also be used to inject drugs intracellularly.

Protocol 10. 'Puffer' drug application

1. Pull a glass microelectrode with a large tip diameter (2–10 μm) (or break back the tip of a finer electrode).
2. Fill the electrode with drug solution at the required concentration, and mount in an electrode holder with side port.
3. Position the electrode 10 to 100 μm from the cell of interest.
4. Eject drug solution on to the cell by applying gas pressure (1–20 p.s.i.) via the side port of the electrode holder. (This is usually controlled by means of a solenoid operated valve.)

4.5.2 Microperfusion
i. The U-tube
This system was first used by Krishtal *et al.* (19) to rapidly apply drugs at different concentrations to the same cell.

Protocol 11. U-tube drug application

1. Form a loop in a length of fine polythene tube (i.d. 0.4 mm, o.d. 1 mm approx.).
2. Pierce a small hole (50–100 μm) at the apex of the loop. This can best be achieved by using a heated insect pin (by attaching it to a soldering iron).
3. Connect one end of the loop via a distribution valve to a series of drug reservoirs, and the other end via a solenoid valve to a peristaltic pump.
4. Position the U-tube in the bath, and adjust the speed of the peristaltic pump so that the flow to waste is slightly in excess of the flow of drug solution from the reservoir, i.e. bathing solution is withdrawn through the whole at a very low rate (50–100 μl/min). On closing the solenoid valve, drug solution will flow out of the hole on to the cell of interest.

Protocol 12. Application of drug using flow pipes

With this technique, solutions of differing ionic composition, or containing different drugs flow into the bath through a series of small inlet tubes. Drug changes are produced by repositioning the recording electrode or the flow pipes so that a different solution flows onto the cell (or patch) which is being recorded from. This technique has been used to study the action of drugs on isolated patches (20).

1. Form flow restrictors by pulling conventional microelectrodes from 1.2-mm diameter glass, and breaking them back to give a tip diameter of 100–150 μm.
2. Cement each flow restrictor into a short length (3–5 mm) of the same glass, using quick-setting epoxy resin.
3. Cement a number of flow pipes together to form a manifold.
4. Connect each flow pipe to a reservoir with a sufficient pressure head to give a flow rate of 0.3–0.6 ml/h.
5. Lower the manifold into the recording chamber.
6. Drug changes are carried out by moving the recording pipette from the mouth of one tube to another.

5. Correlating changes in second-messenger levels with altered ion fluxes

The involvement of second messengers in ligand-induced change in ion fluxes can usually be deduced from a number of simple tests.

(a) If the electrophysiological response is rapid (msec), and observable not only in whole cells but in isolated membrane patches, then the ligand is probably acting directly on the membrane or ion channel without the involvement of intracellular second messengers.

(b) If the ligand causes slow conductance changes, and these effects are dependent upon GTP and mimicked by non-hydrolysable GTP analogues (4, 9), then a receptor-evoked alteration in intracellular second messengers or indirect action on a channel mediated by a G-protein is likely.

(c) If the ligand causes alterations in second-messenger levels with similar kinetics to that found for the alteration in ion flux or electrophysiological response, then direct tests of the second messenger in inducing conductance changes can be carried out. In addition, second message and kinase blockers can be used to deduce the mechanism of the second-messenger action.

Needless to say, complex situations often occur, for example, a ligand (such as acetylcholine) may act both directly and through G-protein-modulated channels. In addition, second messengers may modulate each others actions. Long-term actions of intracellular second messengers can also lead to alterations in gene expression usually through changes in transcriptional regulation that may alter the expression and properties of ion channels. An example of such a phenomenon is the induction of nicotinic acetylcholine receptor expression at motor end-plates by CGRP, probably induced through local changes in cAMP concentrations (22).

Having established the role of a second messenger in a particular ligand-evoked conductance change using cell lines and ion flux or electrophysiological recording techniques, it is then possible to test the existence of the same mechanism in primary neuronal culture, or tissue slices, or even *in vivo*, by means of second-messenger inhibitors and mimetics described in *Table 2*.

References

1. Hille, B. (1984). *Ionic channels of excitable membranes*. Sinauer, Sunderland, Mass.
2. Edelman, G. M., Gall, W. E., and Cowan, W. M. (1987). *Synaptic function*. John Wiley, New York.
3. Wood, J. N., Bevan, S. J., Coote, P. R., Dunn, P. M., Harmar, A., Hogan, P., et al. (1990). *Proc. R. Soc. Lond.*, B **241**, 187–94.
4. Chad, J. and Wheal, H. (1991). *Molecular neurobiology: a practical approach*. IRL Press at Oxford University Press, Oxford.
5. Betz, H. (1990). *Biochemistry,* **29**, 3591.
6. Hollmann, M., O'Shea-Greenfield, A., Rogers, S. W., and Heinemann, S. (1989). *Nature,* **342**, 643.
7. Timpe, L. C., Jan, Y. N., and Jan, L. Y. (1988). *Neuron,* **1**, 659.
8. Nahorski, S. R. (1990). *Transmembrane signalling: intracellular messengers and implications for drug development*. John Wiley, Chichester.
9. Bourne, H. R. (1989). *Nature,* **337**, 504.
10. Burgess, G. M., Mullaney, I., McNeill, M., Dunn, P. M., and Rang, H. P. (1989). *J. Neurosci.,* **9**, 3314–25.
11. Salmon, J. and Flower, R. (1982). *Methods in enzymology*, Vol. 86 (ed. W. Lands and W. Smith), pp. 477–493. Academic Press, New York.
12. Berridge, M. J., Downs, C. P., and Hanley, C. P. (1982). *Biochem. J.,* **206**, 587.
13. Wood, J. N., Winter, J., James, I., Rang, H. P., Yeats, J., and Bevan, S. J. (1988). *J. Neurosci.,* **8**, 3208.
14. Standen, N. B., Gray, P. T. A., and Whitaker, M. J. (1987). *Microelectrode techniques—the Plymouth Workshop Handbook*. The Company of Biologists, London.
15. Changeux, J.-P. (1990). *Trends Pharmacol.,* **11**, 485.
16. Ling, G. and Gerard, R. W. (1949). *J. Cell. Comp. Physiol.,* **34**, 383–96.
17. Gahwiler, B. H. (1988). *Trends Neurosci.,* **11**, 484.
18. Hamill, O. P., Marty, A., Neher, E., Sakmann, B., and Sigworth, F. J. (1981). *Pflügers Arch.,* **391**, 85–100.

19. Krishtal, O. A. and Pidoplichko, V. I. (1980). *Neuroscience,* **5,** 2325–7.
20. Spruce, A. E., Standen, N. B., and Stanfield, P. R. (1987). *J. Physiol.,* **382,** 213–326.
21. Dunn, P. M., Coote, P. R., Wood, J. N., and Rang, H. P. (1991). *Brain Res.,* **545,** 80.
22. Changeux, J.-P. (1991). *New Biologist,* **3,** 413.
23. Latorre, R., Bacigulpo, J., Delgado, R., and Labarca, P. (1991). *J. Bioenerg. Biomemb.,* **23,** 577.
24. Dhalian, R. S., Yau, K. W., Schrader, K. A., and Reed, R. R. (1990). *Nature,* **347,** 184–7.
25. Reuter, H. (1985). *Nature,* **301,** 569.
26. McManus, O. B. (1991). *J. Bioenerg. Biomembr.,* **23,** 537.
27. Baraban, J. M., Snyder, S. H., and Alger, B. E. (1985). *Proc. Natl. Acad. Sci. USA,* **82,** 2538.
28. Buttner, N., Siegelbaum, S. A., and Volterra, A. (1989). *Nature,* **342,** 553.

6

Analysis of protein phosphorylation in cell lines

JAMES R. WOODGETT

1. Introduction

Protein phosphorylation now is regarded as the pre-eminent mechanism of post-translational regulation. Its reversibility, rapid kinetics, and specificity of effect has resulted in its exploitation by most systems of cellular control. Hence, studies of the mechanisms by which cells transduce and respond to signals commonly involve some aspect of this covalent modification. The ability to introduce high specific activity radioisotopes into cultured cells, together with their ease of manipulation and harvest, has made cell lines the system of choice for studying the dynamic processes of protein phosphorylation. In this chapter are described basic techniques for assessing the influence and role of phosphorylation in a given process using cultured cells. Given the increasing realization of the importance and ubiquity of this form of modification, these methods should be of widespread use to researchers wishing to dissect the molecular bases of cellular controls.

2. Metabolic labelling of cells in culture

2.1 Incorporation of ^{32}P

It has been estimated that phosphoproteins comprise up to 50% of cellular proteins. The techniques described here apply to analysis of phosphorylated forms of the hydroxy-amino acids, serine, threonine, and tyrosine. While other phosphoamino acids exist in cells (for example, phosphohistidine) these are generally regarded as enzyme intermediates, have yet to be shown to have a regulatory role, and are extremely labile (and thus not detectable using the methods described below). In normal cells, phosphoserine represents approx. 90% of the total phosphoamino acids, phosphothreonine approx. 10%, and phosphotyrosine between 0.02 and 0.06%. Transformation of cells with certain oncogenes encoding protein-tyrosine kinases or their regulators can elevate phosphotyrosine levels 10- to 20-fold.

The enzymes that catalyse incorporation and removal of phosphate, the protein kinases, and phosphoprotein phosphatases respectively, are generally specific for either serine + threonine or tyrosine (1, 2). [Recently, several examples of protein kinases capable of modifying all three residues have emerged (3).] The existence of competing interconverting enzymes and the sensitivity of transduction pathways to changing conditions ensures that phosphate incorporation into proteins is a dynamic process. Each protein-associated phosphate group may turnover many times per minute and the net level of phosphorylation at a particular residue will depend critically on the ratio of kinase to phosphatase activity. In studying the role of phosphorylation in a process, the aim is thus to take a 'snapshot' of the cell by freezing the interconversion at a given moment. In practice, this is achieved by rapid lysis at low temperature in the presence of inhibitors of the kinases and phosphatases (see Sections 2.2.1 and 4.7).

2.1.1 Hazards of ^{32}P radiation

Since the amounts of proteins isolated from cultured cells are very small, phosphate analysis must be performed using radioisotopes. Virtually all phosphate labelling techniques use the phosphorus-32 (^{32}P) isotope since it emits high-energy β-particles permitting a high sensitivity of detection. This property of ^{32}P also presents significant hazards in its use which must be considered and ameliorated. The amounts of the isotope used in metabolic labelling (10–20 mCi) are significantly greater than used for other common laboratory methods, and require specific consideration. While most research institutions have local guide-lines for working with high-energy radioisotopes such as ^{32}P, there are several additional practical precautions that can substantially reduce exposure.

(a) The range of β-particles in air decreases exponentially with distance. Exposure to the fingers can thus be minimized by using long forceps to move tubes. Likewise, double-sided adhesive tape can be used to fix culture dishes to the bottom of shielded containers. Use of long tubes for centrifugation maintains distance between radioactive pellets and the user.

(b) While two pairs of latex gloves should always be worn to prevent physical contamination, thick rubber kitchen gloves provide some additional protection against irradiation.

(c) 2 cm Perspex is an effective barrier to low levels of ^{32}P (< 1 mCi) but should have an additional layer of lead on the outer surface for larger amounts to act as a barrier for secondary bremsstrahlung radiation. Several Perspex devices for holding tubes and waste are available commercially (for example, Scotlab, Glasgow).

(d) In addition to being harmful to people, metabolic labelling can perturb cells. There is evidence that some cells arrest in S-phase of the cell cycle. As the severity of the effects appear to be somewhat dependent on the cell type, investigators should bear this potential problem in mind when wishing to examine aspects of growth.

2.2 Labelling conditions

Since phosphate is present at millimolar concentrations in normal media, phosphate-free media must be prepared for labelling. For convenience, this can be purchased as a special item from many media companies. Serum should be dialysed against physiological saline (145 mM NaCl, 5 mM KCl, 1.8 mM $CaCl_2$, 0.5 mM $MgCl_2$) to remove phosphate. Obviously, the media requirements for cells in culture vary considerably. As a general rule, labelling media should resemble normal media with the omission of phosphate and serum should be lowered to between 2–4%. The addition of 10 mM Hepes–OH pH 7.4 helps to stabilize pH during the subsequent manipulations. The amount of ^{32}P-phosphate added is dependent on the growth state of the cells and the type of experiment being performed. Resting cells can tolerate higher levels of radioactivity. In most cases 1–2 mCi/ml media are acceptable for labelling cellular proteins.

For labelling of proteins with ^{32}P-phosphate the isotope must be taken up by the cells and incorporated into the γ-phosphate position of ATP. Equilibration of the intracellular ATP pool takes several hours and hence most labelling procedures are performed overnight (16 h). However, shorter-term incubations are possible. For example, 4 h incubation is sufficient to achieve about 90% equilibration. In addition, many cells can survive without serum for this period of time, allowing its omission (and dialysis). Since evaporative losses are low, the media volume can also be reduced if the cells are periodically agitated to prevent drying out. Thus, less label or a higher concentration of label can be used. *Protocol 1* gives a general procedure for ^{32}P-labelling.

Protocol 1. Phosphate-labelling cultured cells

1. Cells for labelling are plated out at least 24 h before the experiment on to suitable dishes (3.5- or 6-cm diameter).

2. 1 h before labelling, remove the medium and wash the cells twice with phosphate-free medium (for example, Tris-buffered saline). Add the labelling medium to the dishes (1 ml for 3.5-cm dish, 3 ml for 6-cm) and place these in the incubator to be used for labelling.

3. Calculate the volume of ^{32}P-phosphate to be added. The half-life of ^{32}P is only 14.3 days, and so its activity must be calculated from the batch reference date. There are several concentrations of ^{32}P-phosphate available: higher concentrations of isotope are preferred since this reduces the volume to be added to the medium. Since the isotope is usually supplied in a solution of dilute HCl, be aware that the addition of large volumes may affect the pH of the media.

4. Add the ^{32}P-phosphate, tilt the dishes to mix and then return these to the incubator. See *Protocol 2* for the lysis procedure on the following day.

2.2.1 Preparation of lysates

There are many types of lysis buffers that have been formulated for particular applications. Conditions for preparation of samples for two-dimensional gels are unique (see *Protocol 4*). For immunoprecipitation of specific proteins, the most important consideration is the type and concentration of detergent. Thus, buffers range from those with no detergent (lysis occurs via homogenization or hypotonic conditions), those with 'gentle' non-ionic detergents (such as Nonidet P-40, Tween 20, etc.) and those with 'harsh' detergents such as sodium dodecyl sulfate and sodium cholate. Since detergents reduce non-specific binding of proteins, the buffers with more detergent are more stringent and generate cleaner immunoprecipitates. However, the stringent buffers will tend to disrupt not only adventitious binding but also complexes of proteins associated with the antigen of interest. Certain antibodies only recognize the antigen in a denatured state. In such cases the lysates must be boiled before use. Thus, the choice of lysis buffer can be critical in influencing the type of proteins to be visualized upon immunoprecipitation. Procedures for the preparation of native cell lysates and denatured lysates are given in *Protocols 2a* and *2b* respectively.

Protocol 2a. Preparation of native cell lysates

1. During the latter period of labelling, add agents to the cell medium (for example, growth factors, phorbol esters) for varying times.
2. Cool the dishes of cells by placement on ice. Remove the medium and wash the dishes twice with ice-cold Tris-buffered saline.
3. After removal of the saline, add 0.5 ml (3.5-cm dish) or 1 ml (6-cm dish) of cold lysis buffer. Dislodge the cells with a cell lifter (Costar) and pipette them into a screw-cap 1.5-ml centrifuge tube (Sarstedt). Tightly cap the tube, vortex, and centrifuge at $15\,000\,g$ for 30 min at 4°C.
4. Following centrifugation, transfer 90% of the supernatant to a new tube. The pellet can be difficult to visualize. If, as occasionally occurs, the pellet is aspirated into the pipette tip, it should be separated into another tube.

Protocol 2b. Preparation of denatured lysates

Steps 1 and 2 as described in *Protocol 2a*.

3. After removal of the saline, add 0.1 ml (3.5-cm dish) or 0.2 ml (6-cm dish) of 0.5% SDS, 10 mM dithiothreitol, 20 mM Tris–HCl pH 7.5. Scrape the cells with a cell lifter and transfer to a screw-cap 1.5-ml tube.
4. Boil for 5 min, cool on ice, and dilute fivefold with RIPA buffer containing no SDS (see formulation in *Table 1*).

5. Centrifuge at 15 000 g for 30 min. Transfer the supernatant to a fresh tube. The pellet is gelatinous and easy to disturb. It can be sucked out with a pipette tip before transfer of the supernatant if preferred.

2.2.2 Optimization of labelling conditions

Since the efficiency of protein labelling is cell-line-dependent and the degree of labelling can vary between dishes, it is important to determine the actual amount of isotope incorporated for each dish of cells. This is done by acid precipitation of protein on to glass-fibre discs (*Protocol 3*).

Protocol 3. Assay of incorporation of isotope into protein

1. Clamp a glass-fibre filter disc (Whatman GF/C, 21 mm dia.) into a suction funnel device (for example, Whatman Cat. No. 950 002).
2. Pre-wash the disc with 2 ml ice-cold 10% TCA.
3. Pipette 2µl of radiolabelled lysate on to the disc. Wash five times with 2 ml of 10% TCA acid followed by 1 ml ethanol (dispose of radioactive washes according to local regulations).
4. Dry the disc in a 50°C oven, place it in a scintillation vial, add enough scintillant to cover the disc and count in a scintillation counter.

Depending on the amount of isotope used, incorporation should be 10^5–10^6 c.p.m. per 2 µl. For example, labelling at 2 mCi/ml, an incorporation of 5×10^5 c.p.m./2 µl should be achieved (representing about 10% overall incorporation of label into protein).

2.3 Biosynthetic labelling of proteins

It is often necessary to label proteins biosynthetically to determine, for example, the optimal conditions for immunoprecipitation or for scanning lines for highest expression of a particular antigen. There are several choices of radioisotope for such labelling using either ^{35}S or ^{14}C. The emission of medium energy β-particles by the former increases sensitivity in conjunction with a scintillant. The additional cost advantage makes ^{35}S-containing amino acids the most commonly used radiolabelled biosynthetic reagents. Either [^{35}S]-methionine or [^{35}S]-cysteine can be used. The former is the usual choice due to the greater relative abundance of methionine in intracellular proteins. However, a mixture of the two is commercially available and an economic alternative (Tran^{35}Slabel, ICN Radiochemicals, about 90% [^{35}S]-methionine, 10% [^{35}S]-cysteine).

Biosynthetic labelling of cells is similar to labelling with ^{32}P-phosphate. However, rather than totally deplete the amino acid used for labelling, it is advisable to retain about 5% of the normal level to prevent starvation responses (add 1/20th volume of normal medium to the methionine- or cysteine-free medium). To take advantage of dual labels (for example, methionine and cysteine), media should be prepared lacking both components.

For normal, equilibrium labelling, 100 μCi/ml of isotope is sufficient. Labelling is usually for a period of 16–18 h. Although the radiation hazard from ^{35}S compounds is insignificant compared with ^{32}P, during incubation some of the isotope is metabolized to a volatile product that can contaminate incubators and their contents. The problem can be substantially reduced by placing the dishes into a sealed box containing a charcoal filter which adsorbs the metabolite. Lysis and estimation of incorporation is exactly as for ^{32}P-phosphate labelling.

3. Analysis of phosphoproteins

There are several strategies available for phosphoprotein analysis which involve either whole-cell analysis or specific protein purification. Since the latter relies on prior knowledge of the protein of interest which may be gained in part from the first approach, it is the whole-cell analysis that will be discussed initially.

3.1 Two-dimensional gel analysis

The complexity of cellular phosphoproteins and their differences in relative abundance and phosphate content demand extremely high resolution separation for analysis. Since many proteins can be resolved by two-dimensional electrophoresis and phosphoproteins compared between different experiments, this analysis can provide valuable information regarding the identification of proteins that respond to particular conditions by changing their phosphorylation state. The actual techniques of isoelectric focusing followed by SDS polyacrylamide gel electrophoresis (PAGE) have been described in detail elsewhere (including a volume of this series (4, 5)). The reader is urged to refer to these sources for the many details that greatly aid in successful separations but which are beyond the scope of this chapter.

3.1.1 Preparation of lysates

Cell lysates for two-dimensional gel analysis can readily be prepared as described in *Protocol 4*.

Protocol 4. Preparation of lysates for two-dimensional gels

Reagents

Staphylococcal nuclease solution:

- 67 μl 3000 U/ml *Staphylococcal aureus* nuclease (Pharmacia)
- 10 μl 2 M Tris–HCl (Sigma)

- 40 μl 2 M Tris base (Sigma)
- 10 μl 1 M CaCl$_2$
- 50 μl 10% Nonidet P-40
- 4.82 ml water
- Store aliquoted at $-70\,°C$.

DNase/RNase solution
- 0.25 ml 2 mg/ml RNase A (Worthington)
- 0.5 ml 2 mg/ml DNase 1 (Worthington DPFF grade)
- 50 μl 1 M MgCl$_2$
- 20 μl 2 M Tris–HCl
- 5 μl 2 M Tris base
- Store aliquoted at $-70\,°C$

2-D sample buffer
- 6 g urea
- 4 ml 10% Nonidet P-40
- 0.5 ml Ampholytes pH 6–8 (Pharmacia/LKB)
- 0.15 g dithiothreitol
- 1.12 ml water
- 10 μl 1% phenol red dye

   Dissolve rapidly at 37°C, filter and store at $-70\,°C$ in aliquots. Do not re-freeze.

Procedure

1. Wash the labelled cells with ice-cold 50 mM Tris–HCl pH 7.5, 1 mM EDTA. For a 3.5-cm dish, add 100 μl of ice-cold *Staphylococcal* nuclease solution, scrape to the edge of the dish with a cell lifter and transfer to 1.5-ml tube.
2. Add 10 μl 3% SDS/10% 2-mercaptoethanol and vortex.
3. Add 20 μl DNase/RNase solution, vortex, place on ice for 1 min, then freeze on dry ice.
4. Repeat for each of the dishes until all are frozen.
5. Lyophilize the samples to dryness.
6. Dissolve each sample in 100 μl 2-D sample buffer. The samples usually need gentle warming to fully dissolve but should not be excessively heated since this promotes carbamylation in the presence of urea breakdown products. If samples are not to be used immediately, they should be snap-frozen and stored at $-70\,°C$.

The isoelectric focusing dimension can either be run to equilibrium or non-equilibrium. The former gives the most easily reproducible gels. The pH gradient used depends on the range of isoelectric points to be resolved. A pH 6–8 Ampholyte mix is a good starting point but other ranges may be required for separation of more acidic proteins. While the dimensions and recommended running conditions for isoelectric focusing gels are highly variable among different manufacturers, the actual gel mix is quite standard:

- 2.75 g urea
- 2 ml 10% Nonidet P-40
- 0.5 ml 30% acrylamide
- 0.675 ml 1% bisacrylamide
- 0.25 ml 40% Ampholytes (appropriate pH range, Pharmacia/LKB)

The mix should be filtered, divided into 1.2 ml aliquots and stored at −70°C. Polymerization is catalysed by the addition of 7 μl 10% ammonium persulfate to an aliquot. Once set, gels are placed in the electrophoresis tank and pre-electrophoresed for 40 min (if equilibrium focusing gels are used). Electrode buffers and voltages vary with the apparatus used. For equilibrium gels, the lower anode buffer is acidic (30 mM phosphoric acid) and the cathode buffer basic (100 mM NaOH). Depending on the dimensions of the tubes, 5–10 μl of sample can be loaded per gel. Following electrophoresis (for example, 1000 V for 14 h), the anode end is marked with ink, the gels extruded from the tubes by air pressure and inserted along the top of a 15% SDS polyacrylamide gel (one SDS gel for each focused gel). The extrusion process can be difficult and several systems have been developed to facilitate this step (for example, use of threaded tubes, Millipore). The second dimension electrophoresis is usually performed at constant voltage for reproducibility. Following electrophoresis, gels are dried and autoradiographed with intensifying screens (see Section 3.3).

Successful, reproducible two-dimensional electrophoresis is something of an art! Only the basic steps have been outlined here to give an idea of the methodology. The interested reader is strongly encouraged to refer to more complete (and lengthy) protocols (4, 5).

3.1.2 Notes regarding electrophoresis of phosphoproteins

Addition of phosphate to a protein results in a net gain in negative charge. Since isoelectric focusing separates proteins based upon their charge, phosphorylated forms of proteins will be displaced towards the anode compared with their non-phosphorylated forms. The effect is additive so that a protein may comprise a series of spots if phosphorylated to differing degrees at multiple sites. In addition, phosphorylation can affect migration in the second dimension, usually causing retardation. The effect is protein- and site-specific and generally occurs by reducing the amount of SDS bound to the protein. In addition to phosphorylation, other post-translational modifications can alter a proteins' charge, confusing the interpretation of data.

Two-dimensional gel analysis using the procedure described here requires that the proteins of interest can be focused and that they are of sufficient abundance for detection. Even if the protein has a suitable isoelectric point, it may co-focus in a region of the pH gradient occupied by several other phosphoproteins. In practice, moderately abundant proteins are most amenable to this procedure. However, important clues to the general effect of a growth factor or other agent can be gleaned from two-dimensional gel separations. Several proteins that respond to a number of factors have been well-characterized, based upon their behaviour in such gels. The induction of phosphorylation of these proteins, including pp42 MAP kinase, p22 and p80 (MARCKS) can give important information regarding the signalling pathways being activated in the cells (see Section 4.5) (6).

3.2 Detection of phosphotyrosine-containing proteins using specific antibodies

Antibodies have successfully been generated that recognize phosphotyrosine. These reagents, which are commercially available (for example, from Amersham, Oncogene Science Inc., ICN), can be used to identify phosphotyrosine-containing proteins either by immunoprecipitation from ^{32}P-phosphate-labelled lysates or by immunoblotting. In the latter case, cells are lysed with Laemmli sample buffer, heated (*Protocol 5*), sonicated to fragment the DNA, and then fractionated on SDS polyacrylamide gels (21). Proteins are electrophoretically transferred to a membrane (for example, Immobilon-P, Millipore) and subjected to immunoblotting with the anti-phosphotyrosine antibody using standard procedures. The only modifications required are the inclusion of 0.1 mM sodium orthovanadate in all buffers following gel electrophoresis (to inhibit phosphatases) and the avoidance of casein to block non-specific binding (casein contains phosphotyrosine). If antibodies are available for a particular protein, immunoprecipitates can be further immunoblotted to test for the presence of phosphotyrosine in the immunopurified protein. While the antibodies available are of high specificity, they do not recognize all phosphotyrosine residues, presumably because of local sequence environments. Similarly, recognition by an anti-phosphotyrosine antibody does not guarantee the presence of this phosphoamino acid since phosphothreonine is sometimes also detected. Thus radioactive phosphoamino acid determination must be used to confirm detection in an immunoprecipitate (see *Protocol 8*). However, these antibodies have proven very useful in the analysis of tyrosine phosphorylated proteins which are usually present at low concentrations in cells.

3.3 Immunopurification

While two-dimensional gel analysis can provide information concerning general effects of an agent on protein phosphorylation, it is usually a starting point for more detailed analysis. If one wishes to determine the effects of a manipulation on the phosphorylation state of a particular protein, the protein must first be isolated from the other cellular proteins. In practice, due to the low amounts of proteins present in lysates derived from only thousands of cells and the need to rapidly isolate the proteins away from phosphatases, such purification is normally achieved by immunoprecipitation. Of course, this depends on the availability of suitable antibodies, but as there are few practical alternatives to immunopurification, it will be assumed that such reagents are part of the investigator's arsenal.

As discussed in the section dealing with lysate preparation (Section 2.2.1), there are several variations for immunoprecipitation which depend on the questions being asked and the properties of the antibodies. The recipes for two commonly used buffers are detailed in *Table 1*. RIPA buffer is relatively stringent, containing low amounts of SDS and sodium cholate. This buffer gives the lowest backgrounds and, if the antibody:antigen complex is stable, is the buffer of choice. The Gentle-Soft buffer contains only non-ionic detergents and should be used for antibodies that are sensitive to SDS. This buffer tends to generate higher backgrounds, although this problem can be reduced by increasing the number of washing steps.

There are several methods which can be used to reduce non-specific background, which can be variable and antibody-dependent. Background is caused primarily by non-specific precipitation of proteins (and RNA) during the incubation with antibody, or by adherence of such components to the added immunoglobulins or protein A used for pelleting the antibody complexes. Radiolabelled RNA can be hydrolysed by the addition of 20 μg of pancreatic RNase to the lysate during the incubation. Pre-clearance of the lysate with protein A preparations can significantly reduce the subsequent precipitation of non-specific proteins. Since it is cheaper and easier to pellet,

Table 1. Commonly used immunoprecipitation buffers

'RIPA'	'Gentle-Soft'
6 mM Na_2HPO_4	20 mM Pipes pH 7.0
4 mM NaH_2PO_4	10 mM NaCl
1% Nonidet P-40	0.5% Nonidet P-40
1% sodium deoxycholate	0.05% 2-mercaptoethanol
1% SDS	0.2 mg/ml Aprotinin
5 mM EDTA	5 mM EDTA
100 mM NaCl	50 mM NaF
50 mM NaF	0.1 mM sodium orthovanadate
0.1 mM sodium orthovanadate	
0.2 mg/ml Aprotinin	

These buffers have been modified to contain EDTA, sodium fluoride, and orthovanadate to inhibit protein kinases and protein phosphatases (see Section 4.7).

pre-clearance is usually performed with formalin-fixed *S. aureus* bacteria expressing protein A on their cell surface (for example, Pansorbin, Calbiochem). Thus lysates are incubated with 100 µl washed *S. aureus* bacteria (resuspended in the appropriate buffer; *Table 1*) for 30 min on ice. The bacteria are then pelleted by centrifugation at 15 000 g for 10 min at 4 °C and the supernatant used for immunoprecipitation.

Precipitation of immunoglobulins is achieved by the addition of immobilized protein A (or protein G) which binds to the invariant Fc domain of certain immunoglobulins. This can be in the form of killed *S. aureus* bacteria (see above) or protein A covalently attached to agarose beads. Since the latter generates lower backgrounds, it is recommended for ^{32}P-phosphate immunoprecipitations. There are several sources including Pharmacia/LKB and Repligen. Preparation simply involves washing the beads with the buffer to be used and storing as a 50% slurry in the presence of 0.02% sodium azide.

Protocol 5. Immunoprecipitation

1. To a 1.5-ml tube on ice add 100–200 µl of pre-cleared lysate, then antibody (1–20 µl depending on concentration) and dilute to 0.5 ml final volume with buffer (*Table 1*). Incubate on ice for 1 h.

2. Add 50 µl of a 50% slurry of protein A agarose and incubate for a further 30 min on a rotary mixer.

3. Transfer the tube contents, using a cut-off pipette tip, to a 15-ml polypropylene screw-cap tube with narrowed end (for example, Falcon 2092), add 5 ml immunoprecipitation buffer and centrifuge for 2 min at 500 g.

4. Carefully remove the supernatant which contains the majority of radioactivity and dispose according to local procedures. Resuspend the pellet in 5 ml immunoprecipitation buffer and re-centrifuge.

5. Repeat step 4 three times. Finally, remove the last of the buffer with a Hamilton-type syringe, resuspend the pellet in 50 µl of 2× Laemmli sample buffer (2% SDS, 0.1 M DTT, 10% glycerol, 0.1% bromophenol blue, 50 mM Tris–HCl pH 6.8) and heat to 95 °C for 5 min. Centrifuge to pellet the agarose beads (5000 g) and pipette off the supernatant for analysis by SDS polyacrylamide electrophoresis.

Procedures for immunoprecipitation are described in detail in *Protocol 5*. The optimal amounts of antibody and protein A agarose added to the immunoprecipitation mix must be determined empirically. This is most conveniently assessed with [^{35}S]-methionine-labelled lysates. Not all of the radioactivity is released from the beads after heating in sample buffer. A further 15–20% of the counts can be recovered by re-extracting the beads following the initial elution. Most of the remaining counts are non-specifically precipitated cellular components. The lysates can be used to precipitate other

Analysis of protein phosphorylation in cell lines

proteins. If this is desirable, the 1.5-ml tube at step 3 should be centrifuged and the supernatant transferred without dilution. It is possible to serially immunoprecipitate several proteins from the same lysate. An advantage is a steady reduction in background. A disadvantage is that antibody binding to protein A is not 100% efficient and so non-precipitated antibodies (with antigen attached) are captured by subsequent additions of protein A, thus complicating the autoradiograph.

Following electrophoresis, gels can be stained with Coomassie blue R250 or dried directly. The most sensitive detection of ^{32}P is obtained using pre-flashed high-sensitivity film (such as Kodak XAR-5), intensifying screens and exposure at $-70°C$ (5).

3.4 Enrichment for phosphotyrosine

Phosphotyrosine is relatively stable to alkaline pH compared with phosphoserine and phosphothreonine. This differential stability can be exploited to increase the proportion of phosphotyrosine detected on gels which is normally present at less than 0.2% of the total phosphoamino acids. Base hydrolysis of proteins can be performed either with polyacrylamide gels (*Protocol 6*) or, following electrophoretic transfer, with PVDF membrane (for example, Immobilon-P, Millipore; see *Protocol 7*). The latter has several advantages including resistance to the distortion caused by alkali treatment of gels.

Protocol 6. Alkali treatment of gels

1. If the gel has been dried (for autoradiography) about 5 mm of gel should be cut from each edge. The gel and backing paper can now be hydrated by submersion in water without shattering (tears sometimes still occur).
2. Transfer the gel to 500 ml 1 M KOH and place it in a 55°C oven for 2 h.
3. Carefully pour off the hot KOH and replace this with 10% acetic acid, 10% propan-2-ol (destaining solution). Change the destaining solution until the gel shrinks back to its original size, rinse with water, dry, and expose to film.

Protocol 7. Alkali treatment of PVDF membranes (7)

1. Following electrophoretic transfer to PVDF, allow the membrane to dry and expose it to X-ray film.
2. After exposure, pre-wet the membrane in methanol and soak it in 50 mM Tris–HCl pH 7.5, 150 mM NaCl for 10 min (the membrane turns a translucent grey).

3. Place the wet membrane in 1 M KOH in a 55°C oven for 2 h. The PVDF filter turns brown.
4. Neutralize the hydrolysed filter for 10 min in 1 M Tris–HCl pH 7.0 then blot it dry and expose to X-ray film.

This technique can be applied to normal SDS gels or to two-dimensional gels, where the detection of rare phosphotyrosyl proteins can be masked by the other, more abundant phosphoproteins. However, the stability of the phosphoamino acids at alkaline pH is relative and is somewhat dependent upon the peptide sequence around the phosphoamino acid. Thus, signals remaining after treatment may contain some residual phosphoserine or, more commonly, phosphothreonine. For unambiguous analysis of phosphoamino composition, partial acid hydrolysis is used.

3.5 Phosphoamino acid determination

The two-dimensional method described below (*Protocol 8*) using two pHs results in complete separation of the three phosphoamino acids and is extremely sensitive with a detection limit of less than 50 c.p.m. in the starting material (8). Usually if a protein can be visualized by autoradiography in a one-week exposure of a gel, it will be amenable to such analysis. However, if many samples require analysis, separation of a line of samples in one dimension is possible, although the results can be ambiguous. The acid hydrolysis is a compromise treatment since under-exposure to acid will result in failure to generate single amino acids whereas over hydrolysis results in breakdown to free phosphate. The starting material is usually a gel slice containing the protein of interest which is eluted, acid precipitated and then hydrolysed. However, it is also possible to hydrolyse proteins that have been electrophoretically transferred to PVDF (see above) which increases the yield of higher molecular mass proteins that are difficult to passively elute from polyacrylamide gels.

Protocol 8. Phosphoamino acid determination from gel samples

1. By alignment with an autoradiographic exposure, the gel piece corresponding to the protein of interest is excised with a razor blade and the backing paper removed.
2. The gel piece is hydrated in 1 ml 50 mM NH_4HCO_3, pH 7.4 and ground into small pieces. This is most easily accomplished with a disposable pestle (Kontes 749520-0000).
3. Add 10 µl 10% SDS and 50 µl 2-mercaptoethanol, tightly cap the tube and boil for 5 min.
4. Incubate on an orbital shaker for at least 2 h (overnight is fine) to elute the protein.

Protocol 8. Continued

5. Centrifuge at 15 000 g for 5 min. Carefully transfer the supernatant to a new tube. Add additional NH_4HCO_3 to the pellet for re-extraction so that the combined supernatants measure 1.2 ml. Agitate the tube for a further 30 min, centrifuge, and then combine the supernatants.

6. Re-centrifuge the entire supernatant and then transfer it to a fresh tube to ensure removal of the gel pieces. Place the supernatant on ice and add 20 µg of RNase A followed by 0.25 ml ice-cold 100% TCA. Incubate on ice for 1 h then centrifuge at 15 000 g at 4°C for 10 min to pellet the protein.

7. Carefully pipette the supernatant into a separate tube. Add 0.5 ml chilled absolute ethanol, invert the tube to mix and re-centrifuge for 10 min. Remove the supernatant and allow the pellet to dry at room temperature for 10 min. The yield of extraction can be monitored by Cerenkov counting and should be 65–70%. If lower, the acid supernatant should be checked for radioactivity. Further incubation of this material on ice may cause additional precipitation and increase the yield.

8. Add 50–100 µl constant boiling HCl (5.7 M) to each tube, tightly cap, and incubate for 1 h in a 110°C oven.

9. Cool the tubes and lyophilize in a centrifugal evaporator (with acid trap). Dissolve the dried amino acids in 5–10 µl pH 1.9 buffer (see *Table 2*) containing 15 parts buffer to 1 part cold phosphoamino acid standards (phosphoserine, phosphothreonine, and phosphotyrosine at 1 mg/ml in water, Sigma). Vortex thoroughly and centrifuge to remove particulate material.

10. Spot up to 4 µl of the hydrolysate on to a cellulose thin-layer electrophoresis plate (such as Merck 5716) in sequential increments of <0.5 µl with drying in between. It is convenient to load up to four samples per plate using the method of Cooper and Hunter (7). The dimensions are shown in *Figure 1*. Also, spot a marker dye comprising 5 mg/ml ε-DNP lysine and 1 mg/ml xylene cyanol FF in 0.5× pH 4.72 buffer (*Table 2*) to monitor the progress of electrophoresis.

11. Wet the plate with pH 1.9 buffer using a blotter of Whatman 3MM paper with 1 cm circles excised at the points of sample loading. The rate of wetting of each sample can be controlled by overlaying a pre-soaked blotter on to the plate and pressing around each circle cut out. All areas of the plate must be wet (but with no puddles).

12. Subject the plate to horizontal electrophoresis at pH 1.9 for 20 min at 1.5 kV. The orientation of anode and cathode is shown in *Figure 1b*. After electrophoresis, allow the plate to air-dry in a fume hood.

13. Wet the plate with pH 3.5 buffer (see *Table 2*) using strips of pre-soaked 3MM paper laid on to the plate parallel with the direction of the first

electrophoresis. Again, selective pressure on the edges of the paper allows control of the migration of the wetting process.

14. Rotate the plate 90 degrees counter-clockwise with respect to the first electrophoresis and electrophorese using pH 3.5 buffer for 16 min at 1.3 kV.

15. Dry the plate in a 65 °C oven. Visualize the phosphoamino acid standards by spraying the plate with 0.25% ninhydrin in acetone and returning the plate to the oven for 15 min. Expose the plate to X-ray film in a paper folder to prevent breakage of the glass plate. Use of luminescent ink or labels allow alignment of the film with the ninhydrin-stained standards.

Phosphamino acids can also be analysed from proteins attached to PVDF membrane (7). In this case, the portion of the membrane containing the phosphoprotein of interest is simply immersed in HCl as in step 8 of *Protocol 8*. After hydrolysis for 60 min, the acid is transferred to a new tube and treated as from step 9.

Table 2. Electrophoresis and chromatography buffers

pH 1.9 buffer	Formic acid (88%)	50 ml
	Acetic acid (glacial)	156 ml
	Water	1794 ml
pH 3.5 buffer	Acetic acid	100 ml
	Pyridine	10 ml
	Water	1890 ml
pH 4.72 buffer:	Butan-1-ol	100 ml
	Pyridine	50 ml
	Acetic acid	50 ml
	Water	1800 ml
pH 8.9 buffer:	Ammonium carbonate	2.0 g
	Water	2000 ml
Chromatography buffer (for phosphopeptides)	Butan-1-ol	750 ml
	Pyridine	500 ml
	Acetic acid	150 ml
	Water	600 ml

3.6 Phosphopeptide mapping

Proteins are often multiply-phosphorylated at distinct sites. The protein kinases and phosphatases acting on these sites may be distinct and thus each site may be regulated autonomously. For this reason it is important to analyse the phosphate content of individual sites. This is usually performed by proteolytic fragmentation of the immunopurified protein, usually with trypsin, followed by separation of the phosphopeptides. The separation can be performed by

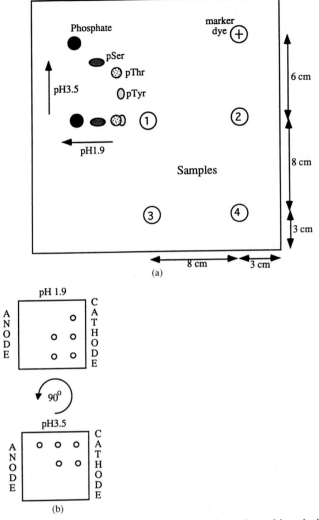

Figure 1. Layout of samples for two-dimensional phosphoamino acid analysis. (a) Up to four samples can be spotted on to the cellulose plate at the indicated positions. The schematic in the upper left sector indicates the migration of the marker phosphoamino acids at the two pHs. (b) After electrophoresis at pH 1.9 the plate is allowed to dry and rotated by 90 degrees anticlockwise for further electrophoresis at pH 3.5.

one-dimensional reverse-phase HPLC or by two-dimensional electrophoresis. The former is simpler but less sensitive. Ideally, the HPLC should have an on-line continuous flow radioactivity detector since this increases sensitivity. Even so, several thousand c.p.m. are usually required for reproducible results depending on the number of phosphopeptides. Electrophoretic separation can yield information from several hundred c.p.m. and is thus often the only feasible method for proteins isolated from labelled cell lines. For this reason, HPLC methodology will not be described in detail here.

However, proteolysed material from step 4 of *Protocol 9* can be injected on to C8 or C18 reverse-phase columns and the peptides eluted with an ascending acetonitrile gradient if required.

3.6.1 Two-dimensional tryptic phosphopeptide mapping

The starting material is usually a gel slice and thus steps 1–7 of *Protocol 8* are followed. Since methionine and cysteine residues can be partially oxidized during the processing, these residues are fully oxidized with performic acid to prevent micro-heterogeneity of peptides. The procedure is given in *Protocol 9*.

Protocol 9. Tryptic phosphopeptide mapping

1. Prepare fresh performic acid by mixing 0.9 ml 98% formic acid with 0.1 ml hydrogen peroxide. Incubate at room temperature for 1 h and chill before use.
2. To the dry pellet from step 7 of *Protocol 8*, add 50 μl of freshly prepared performic acid and incubate on ice for 1 h. (An aliquot can be removed at this stage for phosphoamino acid analysis if required, see step 8 of *Protocol 8*.)
3. Add 0.5 ml water and freeze on dry ice. Lyophilize to dryness in a centrifugal evaporator.
4. Add 50 μl 50 mM NH_4HCO_3 pH 8.3 (1-day-old solutions of NH_4HCO_3 drift to this pH) to the tube and vortex thoroughly. Add 10 μl 1 mg/ml TPCK-treated trypsin (Worthington) and incubate for at least 4 h at 37 °C (overnight is fine).
5. Briefly centrifuge the tube and add a further 10 μl of trypsin and incubate for another 4 h.
6. Add 0.4 ml water, freeze on dry ice, and lyophilize as in step 3.
7. Add 0.3 ml water, vortex thoroughly, and re-lyophilize.
8. Depending on the pH of the intended electrophoresis, resuspend the sample in 0.3 ml of pH 1.9 buffer, pH 4.72 buffer, or water (for pH 8.9 separation) (see *Table 2*), vortex and centrifuge for 10 min at 15 000 g. Transfer the supernatant to a fresh tube and re-lyophilize.
9. Resuspend the peptides in 5–10 μl of the appropriate electrophoresis buffer. Vortex thoroughly, centrifuge to pellet insoluble debris and spot the sample in 0.5-μl aliquots, drying in between each addition, on to cellulose TLC plate (see *Figure 2*). Also, spot 0.5 μl of marker dye (*Protocol 8*, step 10).

Protocol 9. *Continued*

10. Wet the plate with electrophoresis buffer (*Table 2*) using a template made of 3MM paper with a circular hole over the sample.
11. Electrophorese for 20–40 min at 1 kV (depending on the separation obtained). Allow the plate to air dry in a fume hood.
12. Place all dried plates sequentially into a chromatography tank (the larger the better) containing 2 cm depth of chromatography buffer (*Table 2*). Spot an additional 0.5 µl of marker dye 2.5 cm from the bottom for tracking the progress of the chromatography. Lean the plates at an angle against the side of the tank and replace the lid. Leave for 6–8 h until the buffer front is within 3 cm of the top of the plate.
13. Carefully remove the plates and air dry in a fume hood. Mark with fluorescent ink for alignment and expose to X-ray film.

The extensive lyophilization steps are crucial for removal of all salt from the sample before electrophoresis. If residual salt ions remain the peptides will tend to streak. Thus, a good vacuum pump for the centrifugal evaporator

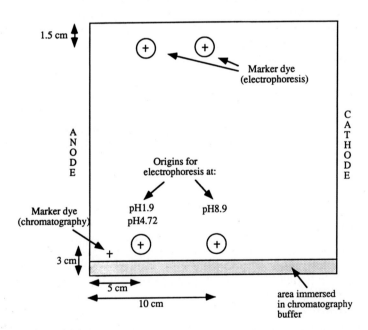

Figure 2. Layout of sample positions for two-dimensional phosphopeptide analysis. The positions of the sample and marker dye is shown for the three different pHs commonly used for electrophoresis. Following electrophoretic separation, the plate is allowed to dry, fresh marker dye spotted, and then immersed into a chromatography tank in the indicated orientation.

is an essential requirement and can substantially speed up the analysis, which, at best, takes 3 days. The choice of electrophoresis buffer is somewhat empirical. With a new protein, all four (*Table 2*) should be tried, and the condition that gives the best separation of phosphopeptides chosen for subsequent analysis. Peptides tend to be easier to solubilize in the lower pH buffers. At least 150–200 c.p.m. should be loaded. Obviously, loading more radioactivity shortens exposure times and is critical for proteins containing many phosphopeptides. It is essential that the chromatography tank be undisturbed during step 12 of *Protocol 9*. For this reason, it is advisable to isolate the tank in a little-used area free from vibration. As for phosphoamino acid analysis, it is possible to analyse proteins electrophoretically transferred to membranes (9). This has the advantage of eliminating the extraction process and is thus faster and more efficient. To prevent irreversible adsorption of the protease (trypsin) to the membrane, after electrophoretic transfer the piece of membrane containing the sample is soaked in 0.5% polyvinylpyrrolidone (Sigma) in 100 mM acetic acid for 30 min at 37°C. After a brief water wash, the sample membrane is ready for protease digestion in ammonium bicarbonate.

There are several commercial sources for thin-layer electrophoresis equipment (such as Pharmacia-LKB). However, one apparatus is custom-designed for separations of phosphopeptides (and phosphoamino acids) and contains a complete description of the use of the equipment for these applications (HTLE-7000 from CBS Scientific, San Diego, CA). The protocols described here require some prior experience with electrophoretic techniques. Novices may find supplemental information in ref. 8.

Phosphopeptide spots can be scraped from plates, the peptides eluted into water, lyophilized, and used for phosphoamino acid determination (*Protocol 8* from step 8) or secondary proteolytic cleavage as an aid to identification (see Section 4.1).

4. Identification of phosphorylation sites: strategies

If the phosphorylation state of a protein is found to change upon stimulation of cells, how can the functional effect of the modification be determined? Identification of the actual site(s) can yield important data concerning the location of the sites within the protein's structural domains and is an essential step for mutagenesis studies. If the amino-acid sequence of the protein is unknown, however, phosphorylation site mapping is confined to low resolution assignment to domains defined by fragmenting the protein into a few large polypeptides by chemical treatment (for example, cyanogen bromide).

Knowledge of the primary sequence, on the other hand, provides a blueprint for phosphorylation site identification. This is attained using a number of approaches that are amenable to analysis of the small number of counts usually recovered by immunoprecipitation from cell lines (<1000 c.p.m.).

4.1 Secondary proteolytic digestion

Trypsin cleaves polypeptides at lysine and arginine residues. Thus, a protein's tryptic peptide composition can be predicted from its amino acid sequence. Knowledge of the type of amino acid targeted can be determined by acid hydrolysis (*Protocol 8*), narrowing the selection of possible candidate phosphopeptides. Experiments must then be designed to discriminate between these possibilities. For example, an eluted tryptic phosphopeptide can be digested with other proteases of known specificity. *Staphylococcus aureus* V8 protease (Boehringer) cleaves at glutamic acid residues in ammonium carbonate buffers, whereas chymotrypsin (Worthington) cleaves after phenylalanine, tyrosine, and tryptophan. Cleavage can be monitored by thin-layer electrophoresis in comparison with untreated peptide. There are several caveats that can complicate tryptic phosphopeptide analysis. If a peptide is phosphorylated at multiple sites it will form a series of spots upon two-dimensional separation. Since trypsin is not an efficient exoprotease, if two arginine and/or lysine residues are adjacent in a sequence, tryptic cleavage will generate two peptides differing only in the presence or absence of an N- or C-terminal basic residue. Tryptic cleavage may also be inhibited by a proximal post-translationally modified residue. Thus, apparently complex tryptic phosphopeptide maps may actually be generated from only one or two phosphorylated peptides (see Section 4.4).

4.2 Manual Edman degradation

Cycles of Edman cleavage can be used to determine the position of a phosphorylated residue from the peptide's amino terminus (*Protocol 10*). However, aliquots must be removed after each degradation cycle and cleavage is not 100% efficient, limiting the effective number of cycles to three or four for samples containing less than about 500 c.p.m.

Protocol 10. Manual Edman degradation

1. Scrape the peptide from the thin-layer plate and elute with water. Centrifuge to remove cellulose and lyophilize the supernatant.
2. Dissolve the peptide in 20 µl water and remove 4 µl as an aliquot of the starting material.
3. Add 20 µl 5% phenylisothiocyanate in pyridine (Pierce). Incubate at 45°C for 30 min.
4. Extract twice with 0.2 ml heptane/ethyl acetate (10:1).
5. Extract twice with 0.2 ml heptane/ethyl acetate (2:1).

6. Lyophilize for 30 min then dissolve in 50 µl trifluoroacetic acid (100%) and incubate at 45°C for 10 min.
7. Lyophilize for 30 min, dissolve in 20 µl water and remove 5 µl sample (1st cycle amino acid). Add back 5 µl water and repeat the procedure from step 3. Remove 7 µl for the next sample, and use all of remaining peptide for the last sample.
8. After three cycles, dry the samples, re-dissolve each in 4 µl pH 1.9 buffer (*Table 2*) and spot each aliquot on to a cellulose thin-layer plate. Electrophorese at 1 kV for 20 min then dry the plate and expose to X-ray film.

An abrupt increase in the amount of phosphate released (running furthest towards the anode) denotes the position of the first phosphorylated residue. Since released phosphate is not purified from the remaining polypeptide at each cycle, once released it is present in all subsequent samples.

Recent developments in covalent attachment for automated liquid phase sequencing (allowing efficient elution of the released phosphate) may allow more detailed analysis, but currently the technique relies on >2000 c.p.m. in the starting material.

4.3 Prediction of migration of phosphopeptides upon electrophoresis and chromatography

Since the migration of a peptide during electrophoresis is dependent upon its charge, if the amino acid composition of a series of peptides is known, their relative migration can be calculated. To estimate an electrophoretic mobility, the expected charges at the pH of electrophoresis are summed, taking into account the amino- and carboxy-terminal charges and the phosphorylated residue (*Table 3*). This charge is divided by the number of residues to give a relative mobility compared with a neutral marker; ε-DNP lysine for pH 4.7 and pH 8.9, phosphotyramine (Sigma), visualized with ninhydrin (*Protocol 8*, step 15) for pH 1.9.

Similarly, the relative chromatographic mobility of a peptide is calculated by summing the constituent R_f values (*Table 3*) and dividing by the number of residues. This generates an estimate relative to the migration of ε-DNP lysine. There are several limitations to this analysis which is thus only intended as a guide. For example, no consideration is made of nearest neighbour effects in which the sequence of amino acids in a peptide may influence its net behaviour. Nevertheless, such predictions often correctly rank peptides relative to each other and provide useful information for estimating the likely identity of a phosphopeptide.

4.4 Use of synthetic peptides

Definitive proof of the identity of a phosphopeptide can be difficult to obtain, especially if the peptide contains several phosphorylated residues and thus

Table 3. Expected amino acid charges and relative mobilities

	pH 1.9	pH 4.7	pH 8.9	Relative mobility[a]
N-terminal NH$_2$ group	+1	+1	+0.5	–
C-terminal COOH group	N	–1	–1	–
Alanine	N	N	N	0.41
Arginine	+1	+1	+1	0.31
Asparagine	N	N	N	0.21
Aspartic acid	N	–0.7	–1	0.22
Cys-SO$_3$H[b]	~–0.1	–1	–1	0.19
Glutamic acid	N	–0.5	–1	0.31
Glutamine	N	N	N	0.29
Glycine	N	N	N	0.3
Histidine	+1	+1	N	0.29
Isoleucine	N	N	N	0.77
Leucine	N	N	N	0.81
Lysine	+1	+1	+1	0.26
Methionine sulfone[b]	N	N	N	0.45
Phenylalanine	N	N	N	0.78
Proline	N	N	N	0.47
Serine	N	N	N	0.33
Phosphoserine	–1	–1	–2	0.20
Threonine	N	N	N	0.41
Phosphothreonine	–1	–1	–2	0.23
Tryptophan	N	N	N	0.69
Tyrosine	N	N	N	0.66
Phosphotyrosine	–1	–1	–2	0.28
Valine	N	N	N	0.62

[a] Relative mobility (R_f) is the distance travelled relative to ε-DNP-lysine (the yellow dye in the marker dye mixture). Values were determined by Dr T. Hunter, Salk Institute, San Diego.
[b] These are the products of performic acid oxidation.

generates several separable spots upon electrophoresis. One approach is to generate mutants in which a suspected residue is engineered to a non-phosphorylatable amino-acid analogue (for example, serine to alanine). However, this requires subsequent introduction of the mutant protein into cells, metabolic labelling, purification, and mapping to show disappearance of a particular phosphopeptide. In addition, some way must be found to discriminate the mutant protein from its endogenous counterpart that will still contain the target residue (for example, use of epitope tags or introduction of a small deletion that allows distinction upon electrophoresis).

A somewhat easier approach is to design a synthetic peptide based upon that of the suspected tryptic peptide; for example, ABCDEXFGY, where X is the serine, threonine, or tyrosine residue. A is preceded by an arginine or lysine, and Y is arginine or lysine. If the peptide is bounded by double basic residues, these should be included in addition to three residues beyond so that the synthetic peptide can subsequently be digested with trypsin to potentially generate multiple forms. Since the peptide must be phosphorylated *in vitro* (see below), sufficient sequence should be present around the target site to allow recognition by a protein kinase. The presence of three residues on either side is usually sufficient. The length of peptide is usually dictated by cost, but most tryptic peptides are, on average, less than 15 amino acids in length.

Once synthesized, the peptide must be phosphorylated. This is easier than might be imagined since, at relatively high concentrations (100 μM), peptides can often be utilized by protein kinases that are not necessarily the physiological kinases in the cell. Thus, many tyrosine kinases are capable of introducing phosphate into just about any tyrosine-containing peptide, albeit inefficiently in many instances.

In the case of serine and threonine residues a little more consideration is needed. Protein kinases exhibit primary sequence preferences in their substrates. Thus, the presence of a basic residue 2–3 residues amino terminal to a serine/threonine will facilitate phosphorylation by cyclic-AMP-dependent protein kinase and calmodulin-dependent protein kinase-II. The presence of a proline residue immediately C-terminal to the serine/threonine residue will generate a target for cdc2 or mitogen-activated protein kinases. Perusal of tables listing preferred phosphorylation sites for particular protein kinases will aid in the choice of kinase for labelling a particular peptide (*Protocol 11*) (10).

An alternative to phosphorylation of peptides following synthesis is to incorporate phosphoamino acids during peptide synthesis. Recently, there has been significant progress in synthesizing peptides containing phosphoamino acids (11). In cases where there are no available protein kinases for phosphorylating a peptide, this may be the only option. However, the synthetic phosphopeptide must still be labelled in some way for detection using a method that does not affect its behaviour upon electrophoresis and chromatography (using a tracer-labelled FMOC amino acid, for example).

Protocol 11. Phosphorylating synthetic peptides

1. Dissolve the peptide in water at a concentration of 1 mM.
2. Prepare a 100 μM solution of γ-^{32}P ATP by mixing 10 μl of 1 mM ATP with 50 μl γ-^{32}P ATP (10 mCi/ml, 3000 Ci/mmol; Amersham Int. Or New England Nuclear) and 40 μl water. Since the half-life of ^{32}P is just 14 days, the nucleotide should be used fresh.
3. To a 1.5-ml tube add 2 μl peptide, 2 μl 40 mM MgCl$_2$, 2 μl 100 mM Hepes–OH pH 7.5, 2 μl 100 μM γ-^{32}P ATP and 10 μl water.
4. Pre-incubate for 2 min at 30°C. Add 2 μl protein kinase, mix and incubate for a further 30 min.

Protocol 11. *Continued*

5. Stop the reaction by addition of 2 μl 100 mM EDTA.
6. Separate the peptide from unincorporated ATP by reverse-phase HPLC [use of C18 Sep-Pak cartridges (Millipore) can save contamination of HPLC equipment].

Once a peptide has been phosphorylated and purified, it can either be spotted directly on to a TLC plate or trypsinized, as described from step 3 onwards of *Protocol 9*. Since two-dimensional separations can be variable, mixtures of *in vivo* tryptic digests and *in vitro* phosphorylated peptides should be subjected to analysis to confirm co-migration.

4.5 Strategies for identification of relevant protein kinases and phosphatases

Once a phosphoprotein has been identified that is sensitive to extracellular stimulation, the mechanism of signal transduction is often of interest. In the case of agonists that operate via increases in cyclic AMP, the mechanism will be through activation of cyclic AMP-dependent protein kinase. Since this enzyme is well characterized and is commercially available (Sigma, Promega Biotec), the question is whether the protein of interest is a direct target for the kinase, or is acted upon by a distinct protein kinase that is regulated by cyclic-AMP-dependent protein kinase. This sort of investigation would benefit from using immunopurified protein as a substrate for an *in vitro* phosphorylation reaction with purified protein kinase (essentially *Protocol 11*) followed by tryptic mapping to ensure the sites labelled in cells were being modified by the protein kinase *in vitro* (*Protocol 9*).

A similar scenario would be presented by known agonists of polyphosphoinositide turnover, in which case the most likely protein kinases would be calcium/calmodulin-dependent kinases or protein kinase C (12). In the latter case, phorbol esters such as phorbol 12-myristate 13-acetate (PMA) act as potent activators and mimic the effect of elevated diacylglycerol. Thus, reconstruction of an effect by PMA is a clear indication of the involvement of protein kinase C. In addition, there are several well-characterized substrates of protein kinase C, such as an 80 kDa protein termed MARCKS (13). Observing an increase in phosphorylation of these 'marker' proteins can provide evidence for activation of particular transduction pathways.

However, many phosphorylation events are not so easy to categorize. For example, an increase in phosphotyrosine occurs on a 42 kDa protein in response to a variety of mitogens and growth factors, including PMA which activates a protein-serine kinase. The 42 kDa protein has recently been identified as a mitogen-activated protein-serine kinase (14). One of the substrates of this enzyme is yet another protein kinase that phosphorylates ribosomal protein S6. Therefore, in studying the 'end-point', S6 phosphorylalation, one is actually monitoring the result of a cascade of interactions between at least three distinct protein kinases.

There are several methods that can be exploited in the identification and subsequent isolation of a protein kinase. Since the protein substrate of interest is usually of very low abundance, the main difficulty occurs in obtaining sufficient quantities of a specific substrate. This might be achieved by bacterial expression of a protein if cloned cDNA is available. There are many suitable bacterial expression vectors for large-scale production of entire proteins or fragments, either alone or as a fusion with a protein that facilitates purification (15, 16). Thus, the protein is expressed, purified and used in a ^{32}P-phosphate incorporation assay as described in *Protocol 10*, with specific phosphorylation being assessed following SDS gel electrophoresis and autoradiography. This approach has been successfully used in the isolation of oncogene-specific protein kinases. A second strategy is to use synthetic peptides. These can be very specific but are more difficult to assay. The addition of two arginine residues to the amino-terminal end of the peptide allows binding to P-11 phosphocellulose paper, which can then be washed free of unincorporated ATP with 10% TCA (15). Peptides can also be separated using high percentage acrylamide gels (25% acrylamide, 0.05% bisacrylamide). In this case, the ion front (just in front of the bromophenol blue dye) is allowed to just run off before the gel is rinsed with water and dried down.

4.6 Assay of protein kinases blotted on to membranes

Some protein kinases can be assayed *in situ* (*Protocol 11*); that is, following SDS PAGE and transfer to a membrane, certain kinases sufficiently renature to be assayed on the membrane (18).

Protocol 11. *In situ* protein kinase assay

1. Prepare the samples from treated dishes of cells by lysis in 0.2 ml of 2× Laemmli sample buffer (*Protocol 5*). Scrape the lysates into 1.5-ml tubes and boil for 2 min.
2. Electrophorese an aliquot (40 μl) of each sample on a standard SDS polyacrylamide gel.
3. Following separation, electrophoretically transfer the proteins to a PVDF membrane (Immobilon-P, Millipore).
4. Following transfer, fully denature the proteins by incubation of the membrane in 7 M guanidine–HCl, 50 mM Tris–HCl pH 8.3, 50 mM dithiothreitol, 2 mM EDTA for 1 h at room temperature.
5. Renature the proteins by incubation of the membrane in 50 mM Tris–HCl pH 8.3, 100 mM NaCl, 2 mM dithiothreitol, 2 mM EDTA, 1% bovine serum albumin, 0.1% Nonidet P-40 for at least 12 h at 4°C.

Protocol 11. *Continued*

6. Incubate the membrane with 30 mM Tris–HCl pH 7.5, 10 mM $MgCl_2$, 2 mM $MnCl_2$, 50 µCi/ml γ-^{32}P ATP (3000 Ci/mmol; Amersham) for 30 min at room temperature.

7. Wash the membrane sequentially in: 30 mM Tris–HCl pH 7.5; 30 mM Tris–HCl pH 7.5, 0.1% Nonidet P-40; 30 mM Tris–HCl pH 7.5, each for 15 min at room temperature.

8. While step 7 removes much of the unincorporated ATP, most of the remaining background can be stripped by a 10 min wash with 1 M KOH at room temperature. This step can distort the recovery of the phosphoamino acids detected, but in some cases is essential.

9. Finally, wash the membrane in water, followed by 10% acetic acid, and then water. Air-dry the membrane and expose it to X-ray film.

The procedure given in *Protocol 11* detects a fraction of protein kinases that are renaturable. The level of incorporation appears to be proportional to their activity state upon lysis and thus can be used to gauge the effect of a particular treatment. The method has the advantage of not requiring metabolic labelling, and reports the molecular masses of the denatured catalytic subunits. Furthermore, phosphoamino acid determinations can be performed on bands excised from the blot (see beginning of *Protocol 8*). A disadvantage is the small number of enzymes that are amenable to renaturation.

4.7 A final word about protein phosphatases

As mentioned in the introduction, protein phosphorylation is a dynamic process. The action of protein kinases is constantly being reversed by protein phosphatases. Thus far, the only consideration for these activities has been to ensure their inhibition during isolation of phosphoproteins. However, their cellular role can be probed, using several useful agents that are potent inhibitors of phosphatases able to pass through the plasma membrane and can therefore be added to live cells. Thus, the dinoflagellate toxin okadaic acid (Gibco-BRL) is a potent inhibitor of protein-serine/threonine phosphatases 2A and 1 (the former exhibiting a tenfold lower K_i (19). Sodium orthovanadate, in contrast, can be used to generally inhibit phosphotyrosyl phosphatases by addition to media (10–40 µM) (20). The effect of addition of these inhibitors is to increase the level of phosphate in proteins in otherwise resting cells. Okadaic acid acts as a non-phorbol ester tumour-promoter, which is believed to be a consequence of it inhibiting dephosphorylation of protein kinase C substrates.

References

1. Hanks, S. K., Quinn, A. M., and Hunter, T. (1988). *Science,* **241,** 42.
2. Alexander, D. R. (1990). *New Biologist,* **2,** 1049.
3. Howell, B. W., Afar, D. E. H., Lew, J., Douville, E. M. J., Icely, P. L. E., Gray, D. A., and Bell, J. C. (1991). *Mol. Cell Biol.,* **11,** 568.
4. Garrels, J. I. (1979). *J. Biol. Chem.,* **254,** 7961.
5. Rickwood, D., Chambers, A. A., and Spragg, S. P. (1990). In *Gel electrophoresis of proteins: a practical approach* (2nd edn) (ed. B. D. Hames and D. Rickwood), pp. 217–43. IRL Press at Oxford University Press, Oxford.
6. Isacke, C. M., Meisenhelder, J., Brown, K. D., Gould, K. L., Gould, S. J., and Hunter, T. (1986). *EMBO J.,* **5,** 2889.
7. Kamps, M. A. and Sefton, B. M. (1989). *Anal. Biochem.,* **176,** 22.
8. Cooper, J. A., Sefton, B. M., and Hunter, T. (1983). *Methods in enzymology,* Vol. 99 (ed. J. B. Corbin and J. G. Hardman), pp. 387–402. Academic Press, Orlando, Florida.
9. Luo, K., Hurley, T. R., and Sefton, B. M. (1990). *Oncogene,* **5,** 921.
10. Kemp, B. E. and Pearson, R. B. (1990). *Trends Biochem. Sci.,* **15,** 342.
11. Zardeneta, G., Chen, D., Weintraub, S. T., and Klebe, R. J. (1990). *Anal. Biochem.,* **190,** 340.
12. Woodgett, J. R., Hunter, T., and Gould, K. L. (1987). In *Cell membranes: methods and reviews* (ed. E. Elson, W. Frazier, and L. Glaser), Vol. 3, pp. 215–340. Plenum, New York.
13. Stumpo, D. J., Graff, J. M., Albert, K. A., Greengard, P., and Blackshear, P. J. (1989). *Proc. Natl. Acad. Sci. USA,* **86,** 4012.
14. Sturgill, T. W., Ray, L. B., Erikson, E., and Maller, J. L. (1988). *Nature,* **334,** 715.
15. Smith, D. B. and Johnson, K. S. (1988). *Gene,* **67,** 31–40.
16. Studier, F. W., Rosenburg, A. H., Dunn, J. J., and Dubendorff, J. W. (1988). *Methods in enzymology,* Vol. 185 (ed. D. V. Goeddel), pp. 60–89. Academic Press, Orlando, Florida.
17. Sahal, D., and Fujita-Yamaguchi, Y. (1987). *Anal. Biochem.,* **167,** 23.
18. Ferrell, J. E., Jr. and Martin, G. S. (1989). *J. Biol. Chem.,* **264,** 20723.
19. Haystead, T. J., Sim, A. T. R., Carling, D., Honner, R. C., Tsukitani, Y., Cohen, P., and Hardie, D. G. (1989). *Nature,* **337,** 78.
20. Klarlund, J. K. (1985). *Cell,* **41,** 707.
21. Laemmli, U. K. (1970). *Nature,* **227,** 680.

7

Isolation and characterization of transfected cell lines expressing neurotransmitter receptors using a calcium dye and flow cytometry

IRA SCHIEREN and AMY B. MacDERMOTT

1. Introduction

An important early application of flow cytometry was to detect subpopulations of cells marked with a fluorescent ligand. This methodology was also used to sort out the targeted cells from the whole population. Traditionally, primary antibodies developed against unique cell markers were coupled to fluorescently tagged secondary antibodies and used to identify subpopulations of cells. In many cases, these markers were cell-type specific for receptors on the plasma membrane. For example, antibodies against immunoglobin receptors specific for subpopulations of lymphocytic T cells were usefully exploited as a way to detect changes in the expression of T cell subsets in different disease states (1). However, since in many cases it has proved difficult to generate antibodies to receptors for hormones, neurotransmitters, and neuromodulators, investigators have been motivated to explore other methods, not dependent on antibodies, with which to identify cells expressing specific receptors on their surface.

One such approach is the use of fluorescent indicators that monitor physiological changes associated with receptor activation. Thus, when DNA-mediated gene transfer is used to express cloned cell-surface receptors, flow cytometry can be used to identify and purify the subpopulation of transfected cells expressing functional receptors. Furthermore, after generating a cell population with high levels of receptor expression, flow cytometry provides a way to characterize the physiological and pharmacological properties of cloned receptors.

The appropriate physiological fluorescent indicator to choose for flow cytometric experiments depends on the receptor under study. For those receptors for which changes in intracellular calcium concentration ($[Ca^{2+}]_i$) is a direct consequence of activation, detection of Ca^{2+} is a useful tool for studying receptor expression. While there are a variety of sensitive fluorescent indicators that may be used to detect changes in $[Ca^{2+}]_i$ (2, 3), the best indicator for this application is indo-1, a dye that is compatible with flow cytometric optics and hardware. Furthermore the light emitted by indo-1 can be detected at two wavelengths and the ratio calculated, giving a value proportional to the absolute $[Ca^{2+}]_i$. The ester form of indo-1 may be loaded into cells by simple diffusion, an important property of dyes meant to be used for experiments on large cell-populations (2, 4). We will describe relevant cellular physiology, hardware and optical considerations, calibration techniques, and approaches to analysis that should be helpful in designing and executing experiments to isolate and characterize cell lines transfected with cDNA clones encoding various types of cell-surface receptors (see also refs 4 and 5).

2. Mechanisms of $[Ca^{2+}]_i$ increase

To permit the most effective experimental design, it is helpful to know the mechanism(s) by which Ca^{2+} is elevated following activation of a particular receptor type. There are three ways in which receptor activation can result in elevation of $[Ca^{2+}]_i$:

- Ca^{2+} entry through receptor-gated channels that are directly permeable to Ca^{2+};
- entry following indirect activation of voltage-gated Ca^{2+} channels resulting from receptor-mediated membrane depolarization; and
- receptor-mediated release of Ca^{2+} from intracellular stores.

If Ca^{2+} entry occurs directly through a channel formed by the receptor under study, cells transfected with a functional form of such a receptor in the simplest case should show an elevation of Ca^{2+} roughly proportional to receptor activation. In neurons, there are several examples of such receptors including the NMDA receptor (6), and one or more of the kainate-sensitive receptors (7, 8). However, there are more numerous examples of receptors in neurons that cause depolarization and subsequent activation of voltage-gated Ca^{2+} channels as the major way of elevating $[Ca^{2+}]_i$. Successful transfection of such receptors into cell lines without voltage-gated Ca^{2+} channels cannot be verified using Ca^{2+} indicator dyes. In principle, this problem could be resolved by co-transfecting the host cell line with a functional cDNA encoding a voltage-gated Ca^{2+} channel (9).

The third mechanism for receptor-mediated elevation of $[Ca^{2+}]_i$ is by release of Ca^{2+} from intracellular stores. These Ca^{2+} release events are mediated by a subclass of receptors belonging to the large superfamily of G-protein-coupled receptors. Known members of this superfamily have an extracellular amino-terminal region, an intracellular carboxy-terminal region, and seven transmembrane spanning regions (10). Following activation, these receptors interact with G-proteins which, in turn, activate a cellular channel or enzyme. The subclass of receptors that release Ca^{2+} interact with G-proteins that activate phospholipase C, an enzyme that produces the intracellular second messengers inositol 1,4,5-trisphosphate ($InsP_3$) and diacylglycerol (11). One of the primary targets of $InsP_3$ is the $InsP_3$ receptor located in the membrane of the endoplasmic reticulum. Activation of this receptor results in release of intracellular Ca^{2+} from the endoplasmic reticulum storage pool. Changes in $[Ca^{2+}]_i$ following activation of these G-protein-coupled receptors are often transient, particularly when they are exclusively dependent on release of intracellular Ca^{2+}. One diagnostic for this type of response is that it can be maintained in the absence of extracellular Ca^{2+}.

3. Calcium indicator

Indo-1 is a fluorescent dye that exhibits a shift in fluorescence emission wavelength when it binds Ca^{2+} (2). It is available in an esterified form that easily passes through the cell membrane into the cytoplasm (see *Protocol 1*), where it can be cleaved by esterases to produce a membrane-impermeant, Ca^{2+}-sensitive form. Roger Tsien has developed a family of such dyes, including indo-1 and the better known fura-2 (2). Compared to dyes that simply increase fluorescence in response to Ca^{2+}-binding, fura-2 and indo-1 are advantageous for obtaining quantitative measurements of the intracellular concentration of Ca^{2+} since the wavelength of peak fluorescence shifts in the presence of Ca^{2+}. Specifically for indo-1, at low Ca^{2+} levels, emission is maximal at 480 nm and minimal at 405 nm. As Ca^{2+} concentration rises the emission curve shifts and becomes maximal at 405 nm.

Good choices for excitation wavelengths for indo-1 are 355 and 363 nm since indo-1 undergoes changes in fluorescence intensity at 405 and 480 nm of similar amplitude but opposite direction when excited at these wavelengths (*Figure 1A* and *1B*). Even though the shifts in the emission wavelengths with changes in $[Ca^{2+}]$ at other excitation wavelengths are not as symmetrical, a useful signal can be produced. Thus it is possible to use the ultraviolet (UV) optics of argon or krypton lasers that generate peak output at 363 nm and 355 nm respectively, to effectively excite indo-1. The helium–cadmium (He–Cd) laser peaks at 325 nm. Excitation at this wavelength is more problematic since the change associated with Ca^{2+} at wavelengths near 480 nm is minimal (*Figure 1C*). Under such conditions, the fluorescence will still be proportional to $[Ca^{2+}]$ but may not provide estimates of absolute $[Ca^{2+}]$. *Figure 1* shows the emission spectra of indo-1 with Ca^{2+} concentrations of 50 nM and 1 μM when excited with three different excitation wavelengths: (A) 363 nm, (B) 355 nm, and (C) 325 nm.

Ratio measurements have several advantages over single wavelength measurements. When only one emission wavelength is used to monitor $[Ca^{2+}]_i$,

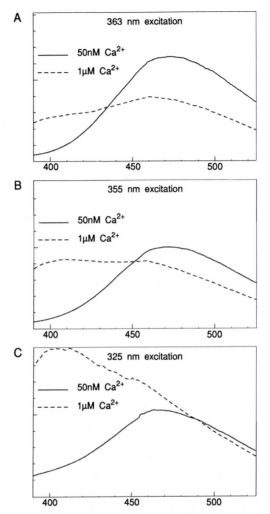

Figure 1. The emission spectra of indo-1 acid in 50 nM Ca^{2+} and 1 µM Ca^{2+} solutions with excitation at 363 nm in panel **A**, 355 nm in panel **B**, and 325 nm in panel **C**. These excitation wavelengths correspond to the outputs of argon, krypton, and He–Cd lasers, respectively. These spectra, recorded on the SLM Aminco 5000C fluorimeter, are uncorrected.

it may be difficult to distinguish a cell with high $[Ca^{2+}]_i$ from a cell loaded with more dye. This can be a problem since cells in a heterogeneous population may not load esterified dyes uniformly well. Even in a homogeneous cell population there are differences in cell size and esterase content that affect how much of the Ca^{2+}-sensitive dye is trapped within each cell. Using ratio measurement obviates this problem. Most flow cytometers have hardware that calculates the ratio of the signal from two photodetectors in real time.

For systems that do not have this on-line capability, ratios must be calculated after data collection.

In practice, we have been collecting fluorescence emission at 400 and 500 nm (see Section 4.1). When the ratio of the 400 and 500 nm emissions is calculated, the resulting signal is non-linearly proportional to the absolute $[Ca^{2+}]_i$. Therefore it is possible to calibrate the system and measure actual $[Ca^{2+}]_i$ if desired. The ratio measurement also provides improved signal detection over noise, thus making the system sensitive to smaller shifts in Ca^{2+} levels than is possible with a single wavelength measurement. Finally, working with ratios will also compensate for instability in the laser excitation beam. This opens up the possibility of using small, inexpensive, low-powered He–Cd lasers which might otherwise not be stable enough for high-resolution flow cytometry. Even though the emission at 500 nm is not very Ca^{2+}-sensitive with 325 nm excitation, it is sensitive to the presence of the dye and thus the 400 nm/500 nm emission ratio will nullify the excitation instability.

Protocol 1. Loading cells with indo-1 ester

A simple protocol for loading cells with indo-1 is given below (see also refs 4 and 12).

1. Remove adherent cells from culture dish or flask with brief incubation at 37°C with trypsin–EDTA or, if possible, an enzyme-free cell lifting solution (Chapter 1). Trypsin should be avoided if possible as it may proteolyse the receptors to be studied. Cells should be fully dissociated to facilitate their smooth and rapid flow through the cytometer. If large aggregates of cells are present, they may be separated from the single cells by letting the cell suspension sit for a few minutes. The large aggregates will settle to the bottom of the tube and the single-cell suspension can be pipetted off.

2. Load the cell suspension with dye by incubation at room temperature in a solution consisting of 10 μM indo-1 acetoxymethyl ester (Molecular Probes) in Hepes-buffered Hanks' saline with the addition of 2 mM Ca_2Cl, 0.1% BSA and 0.1% glucose. Loading time can range from 10 to 20 min but should be kept as brief as possible. This prevents excessive loading that may result in dye entering intracellular membrane-bound compartments and distortion of the experimental results (12, 13). Optimum loading time will vary somewhat for different cell lines as it is dependent on the intracellular esterase content of the cells.

3. After indo-1 loading, wash the cells in the enriched Hanks' saline and resuspend at 1×10^6 per millilitre. Some cell lines increase their $[Ca^{2+}]_i$ as a result of the trauma associated with these procedures. An additional 15 min incubation at room temperature can bring Ca^{2+} levels down. This post-loading incubation period also allows for more complete de-esterification of the intracellular dye. Some cell lines will actively excrete the indo-1

Protocol 1. *Continued*

dye slowly after loading (12). This can be monitored by watching the fluorescence emission at one wavelength during the course of the experiment. Dye excretion can be slowed by maintaining the cells on ice. However, the cold temperature may slow down physiological processes in the cells and decrease or eliminate the $[Ca^{2+}]_i$ shift in response to neurotransmitter. Furthermore, ratio measurement minimizes the effect of dye loss on estimates of $[Ca^{2+}]_i$.

---

4. Flow cytometry

4.1 Hardware

4.1.1 Optics and electronics

Indo-1-loaded cells can be excited with an argon or krypton ion laser, emitting 100 mW at wavelengths of 363 and 355 nm, respectively. It is also possible to excite indo-1 with the output of a small air cooled He–Cd laser. These relatively inexpensive, long-lived, low-power lasers can be easily retrofitted to a flow cytometer that does not have UV capability. The peak emission of a He–Cd laser tuned for UV operation is around 325 nm with a power output of 13 mW, which is sufficient for Ca^{2+} measurement.

In order to detect and ratio indo-1 signals, the flow cytometer should be set up with filters comparable to those illustrated in *Figure 2* (5). The fluorescent emission is split with a 448 nm dichroic long-pass filter (Andover). The shorter wavelengths are reflected to a 395.5–405.5 nm narrow band pass filter (Oriel) in front of photomultiplier tube (PMT) #1. The longer wavelengths pass through the dichroic filter and intersect a 500 nm wide bandpass filter constructed by placing a 530 nm short-pass filter (Coulter) in front of a 485 nm long-pass filter (Coulter). This combination of filters is placed in front of PMT #2. This arrangement of narrow band filtering for the numerator and wide band filtering for the denominator of the ratio provides the best separation of low and high Ca^{2+} populations for this system. The PMTs convert the light they collect to a proportional voltage in the range of 0 to 10 V. The ratio circuitry or data acquisition computer takes the output of PMT #1 for one cell and divides it by PMT #2 signal for the same cell and multiplies it by 10. This results in an output signal between 0 and 10 V that can be processed by the flow cytometer.

The sensitivity of each PMT is partially determined by the voltage provided it. Precautions must be taken when setting up experiments to optimize the voltage settings on the PMTs to produce ratios within the 0 to 10 V range. If the 400 nm signal is used as the numerator for generating a ratio, PMT voltages should be set to provide the smallest possible output signal while still providing a minimum 1 V signal at low $[Ca^{2+}]_i$. Under these conditions, the

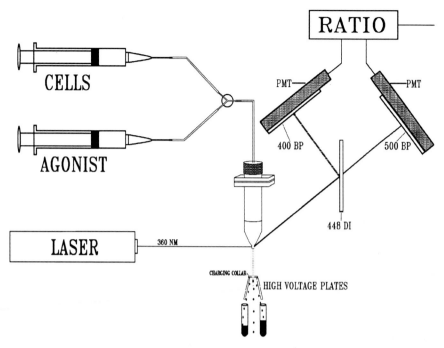

Figure 2. The optical set-up necessary to detect changes in $[Ca^{2+}]_i$ with indo-1 loaded cells and the dual syringe injector system for maintaining peak transient $[Ca^{2+}]_i$ responses at the laser intersection. (Taken from ref. 5.)

500 nm signal should be maximized, with less than 10% of the signal going off scale (higher than 10 V).

4.1.2 Sample delivery system

To measure changes in $[Ca^{2+}]_i$ as a function of time, the standard sample delivery system of the flow cytometer may be used. The standard sample delivery system on most flow cytometers consists of a sample vial and rubber stopper with two openings. One opening connects to a regulated nitrogen source to pressurize the vial. The other opening allows the sample to exit through silicon tubing from the pressurized vial. The differential pressure between the sample and sheath stream regulate the sample flow rate. To measure agonist-induced changes in $[Ca^{2+}]_i$, the top of the sample vial can be removed and agonist added. Analysis and sorting is possible only if $[Ca^{2+}]_i$ remains elevated for an extended period of time.

It is helpful to have information about the time-course of the Ca^{2+} response following receptor activation in order to choose the best form of drug application. For example, if the response shows rapid inactivation or desensitization, then continuous and prolonged exposure to agonists, as occurs with the standard sample delivery system, may result in decrease or loss of the response. This may result in a false-negative assessment of receptor expression since the response might end before it can be detected. Although desensitization can occur with activation of any of the three mechanisms of $[Ca^{2+}]_i$ elevation, it occurs most typically with responses involving release of Ca^{2+} from intracellular stores.

To measure and sort cells that show only transient changes in $[Ca^{2+}]_i$, the sample delivery system of the flow cytometer can be modified (5). The pressurized vial system is replaced with two syringe injectors and a 'Y' connector. The Y connector may be made by imbedding three 15-mm lengths of 26-gauge stainless steel tubing in a piece of Plexiglass such that they intersect at a centre compartment. Silicon tubing of 0.25 mm i.d. is used to connect the syringes to the Y connector and the Y connector to the flow cell. This allows the drugs and cells to be kept apart until mixing at the centre compartment (see *Figure 2*). The pressure on the syringes must be identical to achieve good mixing and even flow of cells. The length of tubing from the exit of the Y to the flow cell, together with the flow rate of the solutions, controls the duration of agonist application. This can be easily varied from less than 1 sec to many seconds. One can empirically determine the appropriate length of tubing that will put the cells in front of the laser and detectors near the peak of their transient change in $[Ca^{2+}]_i$.

4.2 Data display and analysis

Time plays an important role in experiments that measure ongoing physiological changes in cells. Since a flow cytometer makes a ratio measurement on one cell over an approximately 3-μsec period, a biological response lasting many seconds cannot be measured in one cell. However, the time-course can be monitored by using the change in $[Ca^{2+}]_i$ over time within the population of cells using the standard sample delivery system. When the whole sample of cells is exposed to drug and the response followed over time, the results can be displayed in a dual parameter histogram in real time during the experiment. An example of such a graph is shown in *Figure 3A* and *B*. These histograms show time progressing on the *x*-axis, $[Ca^{2+}]_i$ (400 nm/500 nm) on the *y*-axis and cell number on the *z*-axis, with higher cell concentrations represented by darker areas.

In the experiment illustrated in *Figure 3A*, a baseline Ca^{2+} level was

---

Figure 3. The human astrocytic cell line responds to 100 μM MCh by an elevation of $[Ca^{2+}]_i$ that is dependent on both Ca^{2+} release and Ca^{2+} entry. **A**: A dual parameter histogram showing the distributed population response to 100 μM MCh over time. The density of the histogram corresponds with cell number. At 6.7 min, 5 mM EGTA was added to the cell/solution mixture, dropping the $[Ca^{2+}]_i$ of the whole population to somewhat lower levels than before MCh was added. **B**: Dual parameter histogram showing the response to 100 μM MCh in the continuous presence of 10 mM EGTA. The transient response is still

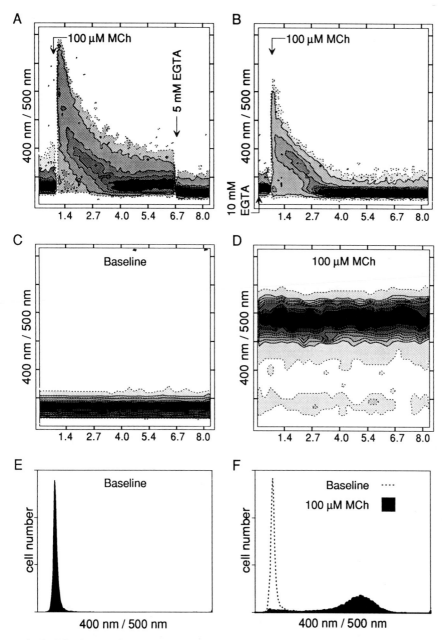

evoked while the steady-state response is blocked under these conditions. **C**: Dual parameter histogram of the same cells running through the dual syringe injector system in the absence; and **D**: the presence of 100 μM MCh. **E**: A single parameter histogram showing the fluorescence ratio data vs. cell number in **C** and **F**, showing the fluorescence ratio data vs. cell number in **C** (*dotted line*) and **D**.

recorded from a human astrocytic cell line (14), before agonist was added. At the indicated time, flow was stopped and 1 mM acetyl-beta-methylcholine (MCh) was added to the sample tube to achieve a final concentration of 100 μM. By the time cell flow was re-established, the $[Ca^{2+}]_i$ in the population of cells appeared to be already decreasing, suggesting that the peak of the Ca^{2+} transient occurred earlier. By 5.4 min after the start of the run, the average population response decreased to a steady-state level that was slightly higher than the resting distribution of $[Ca^{2+}]_i$. Notice that the distribution of $[Ca^{2+}]_i$ within the agonist-activated population overlaps with the resting distribution of $[Ca^{2+}]_i$. Thus, even though this is a population assay, it is more informative in this regard than a time course obtained with a fluorimeter. With a flow cytometer, the population distribution of $[Ca^{2+}]_i$ is provided by making rapid measurements on individual cells within the population as a function of time, rather than averaging the whole population response.

With the dual syringe injector system, data collection can occur near the time of peak $[Ca^{2+}]_i$ elevation. When such an experiment is done with an astrocytic cell line and viewed using a time vs. $[Ca^{2+}]_i$ histogram, the baseline or resting distribution of $[Ca^{2+}]_i$ (*Figure 3C*) looks similar to that recorded just prior to drug application in *Figure 3A*. However, when the cells are allowed to mix with MCh for a fixed period of time, the response is robust in that most of the cells have elevated $[Ca^{2+}]_i$ at that time (*Figure 3D*). Furthermore, the time of agonist exposure can be estimated by comparison of response amplitude (*Figure 3D*) with the time after MCh application at which the response amplitude is similar (*Figure 3A*). If we assume that the response with the syringe injector system is occurring on the falling phase of the response to MCh, then this occurs at the 67-sec time-point, which corresponds to about 17 sec of exposure to MCh. Since all of the cells are recorded with uniform time of drug exposure, these data can be viewed and readily analysed in a single parameter histogram in which time is not a variable. The single parameter histogram of resting $[Ca^{2+}]_i$ for the data in *Figure 3C* is shown in *Figure 3E*. The MCh-activated response is shown in *Figure 3F* with the baseline histogram overlaid for comparison. These histograms can be subtracted to obtain the percentage of cells that show $[Ca^{2+}]_i$ higher than the resting distribution of $[Ca^{2+}]_i$. They can be analysed for mean or median response, and for maximum and minimum response amplitude. These histograms also provide a convenient way to set the sorting criteria for purifying receptor-positive cells. The dual syringe injector system sustains cell flow at an optimal response amplitude, enabling long sorts, detection of small subpopulations, and more accurate analysis when transient $[Ca^{2+}]_i$ shifts are involved. (See ref. 5 for an analysis of the accuracy of subpopulation detection with this method.)

4.3 Calibration

The addition of a Ca^{2+} ionophore in the presence of normal Ca^{2+} concentration in the extracellular media (about 2 mM) can cause $[Ca^{2+}]_i$ to rise to the extracellular concentration. The addition of an excess of EGTA, the Ca^{2+} chelator, will decrease extracellular $[Ca^{2+}]$ to below 10^{-7} M and $[Ca^{2+}]_i$ should follow. These manipulations provide the ratio measurement at the maximum and minimum $[Ca^{2+}]$ that can be detected with indo-1. It is possible to estimate $[Ca^{2+}]_i$ levels anywhere in between these values based on the ratio measurement and also to calculate $[Ca^{2+}]_i$ changes in response to agonist activation. The relationship between the ratio fluorescence measurements and $[Ca^{2+}]_i$ is represented by the equation (2):

$$[Ca^{2+}]_i = K_d \times \frac{(R - R_{min})}{(R_{max} - R)} \times \frac{F_{500\ free}}{F_{500\ bound}} \quad [1]$$

where R is the fluorescence at 400 nm/500 nm of the unknown $[Ca^{2+}]_i$, R_{min} is the fluorescence ratio of cells exposed to ionophore and EGTA, and R_{max} is the fluorescence ratio of cells exposed to ionophore in the presence of a saturating $[Ca^{2+}]$. $F_{500\ free}$ is the average 500 nm fluorescence in the presence of ionophore and EGTA and $F_{500\ bound}$ is the average 500 nm fluorescence from cells in the presence of ionophore and high Ca^{2+}. The indo-1/Ca^{2+} dissociation constant (K_d) estimated for mammalian cell intracellular salt composition is 250 nM (4). Useful ionophores for this application are ionomycin and 4-bromo-A23187 (see ref. 15 for discussion). *Figure 4* is a two parameter histogram showing $[Ca^{2+}]_i$ changes over time with the addition of 20 μM ionomycin and then EGTA. Data from this histogram can be entered into equation 1 to determine the resting levels of $[Ca^{2+}]_i$ and $[Ca^{2+}]_i$ in response to neurotransmitter receptor activation.

It is often difficult to obtain accurate *in vivo* calibration of intracellular indicator dyes and the degree of difficulty varies with cell type. The main problem is the resistance of the normal cellular Ca^{2+} regulatory mechanisms to prolonged elevation of $[Ca^{2+}]_i$. These mechanisms can result in long and variable time-delays before $[Ca^{2+}]_i$ reaches extracellular Ca^{2+} levels in the presence of Ca^{2+} ionophore, if it can at all. The Ca^{2+} permeability created by the ionophore must be great enough to overcome normal homeostatic mechanisms in order for the calibration technique to work properly. For instance, many cell types have the ability to extrude Ca^{2+} during periods of $[Ca^{2+}]_i$ elevation via Ca^{2+}-ATPase pumps or Na^+–Ca^{2+} exchanger. Conditions can be improved by dissipating the Na^+ gradient with the Na^+ ionophore, monesin. Since Na^+/K^+-ATPase may be highly active, a K^+ ionophore, nigercin (Molecular Probes) and a pump inhibitor, ouabain, can be used to reduce the pump ability to restore Na^+ levels and Ca^{2+} pumping activity. The ideal combination and concentration of ionophores will vary with cell type. These issues are presented and discussed in detail by Chused *et al.* (4) and Williams and Fay (15).

Another innovative method for calibrating ratio dyes has been described by Kachel *et al.* (16). This method makes use of a removable motorized rotating disc with two 5-mm diameter holes placed at opposite ends of the disc. This

Isolation and characterization of transfected cell lines

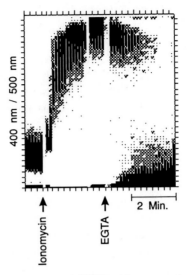

Figure 4. The two parameter, time vs. $[Ca^{2+}]_i$, histogram depicts the changes in the 400 nm/500 nm signal when various reagents are added to a suspension of the AR42J pancreatic acinar cell line. The density of the histogram corresponds with cell number. The actual $[Ca^{2+}]_i$ can be determined by equilibrating $[Ca^{2+}]_i$ with $[Ca^{2+}]_o$ (2 mM) following application of 20 µM ionomycin to the cell suspension. An approximation of 0 $[Ca^{2+}]_i$ level can be achieved by applying EGTA to the cell suspension. The ratio values and 500 nm values under these conditions can be put in equation 1 to determine the resting level of $[Ca^{2+}]_i$. (Taken from ref. 5.)

disc is placed in front of the laser excitation beam and rotated at 3000 r.p.m. The periodic interruption of the excitation beam creates a pulsed signal necessary to trigger the flow cytometer to collect data. If different known concentrations of Ca^{2+} solutions are run through the flow cell with indo-1 acid, data can be collected in the same way as with indo-1 loaded cells. The resulting histograms provide a scale of various Ca^{2+} concentrations that can be used to calibrate data with cells. This method can provide quick and accurate calibrations for experiments. The main difficulty is in making accurate standards of known concentrations of Ca^{2+}. A procedure for this was provided by D. Reichling (see also refs 15 and 17) and is given in *Protocol 2*.

Protocol 2. Preparing standardized $[Ca^{2+}]$ buffer

1. Make up a series of solutions with strongly buffered $[Ca^{2+}]$ and having ionic compositions similar to that of the cellular environment in which $[Ca^{2+}]$ will be measured for use as Ca^{2+} standards. The approximate desired $[Ca^{2+}]$ are achieved by mixing differing proportions of two solutions: one containing EGTA with no Ca^{2+}, and the other containing equimolar $CaCl_2$ and EGTA (18). Both solutions contain (in millimolar) 60 N-tris(hydroxymethyl)-2-aminoethane sulfonic acid (Tes), 120 KCl, 20 NaCl, and 20 EGTA. The EGTA solution contains 1.98 $MgCl_2$ and the Ca:EGTA solution contains 1.06 $MgCl_2$ and 20 $CaCl_2$. The proportion of the two solutions to be used should be varied over the entire range of possible combinations. The free Ca^{2+} concentration ($[Ca^{2+}]$) is determined from equation 2 (19).

$$a \times [EGTA]_{tot} = \frac{[EGTA]_{tot} \times Ca\text{-}K^{EGTA}[Ca^{2+}]}{1 + Ca\text{-}K^{EGTA}[Ca^{2+}] + Mg\text{-}K^{EGTA}[Mg^{2+}]} + \frac{[Tes]_{tot} \times Ca\text{-}K^{Tes}[Ca^{2+}]}{1 + Ca\text{-}K^{Tes}[Ca^{2+}] + Mg\text{-}K^{Tes}[Mg^{2+}]} \quad [2]$$

where $Ca\text{-}K^{EGTA} = 6 \times 10^6\ M^{-1}$, $Mg\text{-}K^{EGTA} = 46\ M^{-1}$, $Ca\text{-}K^{Tes} = 1\ M^{-1}$, and $Mg\text{-}K^{Tes} = 1\ M^{-1}$ (19). The series of Ca^{2+} solutions generated by mixing can be precisely standardized in the stages outlined below.

2. Titrate a sample of each newly generated standard solution with $CaCl_2$ while monitoring pH to determine the $[EGTA]_{free}$ (20). $CaCl_2$ should be added incrementally to the solution until the pH has clearly stopped decreasing. The exact point at which all the free EGTA has been titrated is determined from the V-shaped plot of volume of $CaCl_2$ added vs. pH. At this point the total moles of $CaCl_2$ added equals the $[EGTA]_{free}$. The starting pH of the titrated solution should be adjusted using KOH so that the titration end-point occurs at a pH greater than 7.5. This ensures that all free EGTA is in the form $EGTA^+$, rather than HEGTA, giving the plot a steeper slope and a sharper end-point.

3. Next, titrate each solution to determine $[EGTA]_{total}$ as above, but use $CdCl_2$ which displaces both H^+ and Ca^{2+} that is bound to EGTA (21). In this case the starting pH is adjusted so that the titration end-point occurs in the range pH = 4–4.5 in order to maximize the apparent affinity of EGTA for Cd^{2+} while minimizing the apparent affinity of EGTA for Ca^{2+}.

4. These measured values are then used in equation 3 to calculate the value of a:

$$a = 1 - [EGTA]_{free}/[EGTA]_{tot}. \quad [3]$$

5. Then substitute this value of a (and $[EGTA]_{total}$) into equation 2 to calculate $[Ca^{2+}]$. (Rather than solve this equation, it is easier to construct a 'look-up' table of a vs. $[Ca^{2+}]$ for a given $[EGTA]_{total}$.) A similar approach is described by Tsien and Pozzan (17) in which pH titration is used to measure precisely the concentrations of EGTA and Ca^{2+} in the starting solutions then calculating the final free Ca^{2+} following mixing to generate a set of standards.

5. Application
5.1 Identification and selection of cells with functional receptors

Transfection of a putative cDNA for a neurotransmitter receptor into a cell line provides a system for determining whether the cDNA clone under study is functional. Flow cytometry can be used to assay receptor function; a positive physiological signal following receptor activation can be used to identify and purify receptor-expressing cells. Recent experiments in which several subtypes of the serotonin, or 5-hydroxytryptamine (5-HT), receptor family were transfected into a host cell-line demonstrate the usefulness of the indo-1 system in identifying a unique subpopulation of cells with neurotransmitter receptors on their surface.

In these experiments, a cDNA insert for the 5-HT1c receptor was initially shown to be a functional clone by expression in *Xenopus* oocytes. The cDNA was then subcloned into the mammalian cell expression vector pMV7 (22, 23). This vector contains a murine leukaemia virus long terminal repeat that serves as a promoter for the expression of the 5-HT1c receptor cDNA in mammalian cells, as well as an independent expression cassette encoding neomycin phosphotransferase allowing for selection of stable, neomycin-resistant transfectants (24). The host cells used in the transfection were the NIH-3T3 mouse fibroblasts. Stably transfected cells, selected by resistance to the neomycin analogue, G418, were isolated and loaded with indo-1 to test for receptor expression (22).

The initial test for functional receptor expression on the cell sorter was performed using the dual syringe injector technique, with the fluorescence from each cell being detected after an estimated 8-sec exposure to 5-HT. As shown in *Figure 5A*, approximately 30% of the neomycin-resistant cells responded to 5-HT exposure with a significant increase in $[Ca^{2+}]_i$. This indicates that neomycin resistance functions only as a partial selection for successful expression of an independent gene and does not ensure that cells containing the transfected plasmid express the cloned cDNA at appreciable levels. Cells with high $[Ca^{2+}]_i$ were identified in a single parameter histogram, gates were placed around the high $[Ca^{2+}]_i$ peak to define the sort window, and the response-positive cells were sorted. The positive cells were plated, grown, and analysed one to two weeks later. In this enriched population, 85% of the cells were found to respond to 5-HT with a large increase in $[Ca^{2+}]_i$ (*Figure 5B*). The ability of the cell sorter to select a subpopulation of cells responding to 5-HT allows for a one-step enrichment of a stable cell line expressing relatively high levels of 5-HT receptors.

Following isolation of a cDNA for the 5-HT2 receptor, flow cytometric detection of Ca^{2+} response was the only functional assay used for detailed functional characterization (25). In these experiments, response to 5-HT was

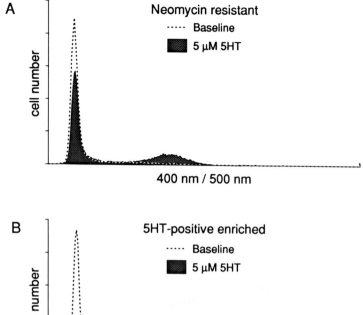

Figure 5. The histograms show the $[Ca^{2+}]_i$ response of NIH-3T3 cells transfected with the 5-HT1c receptor. **A**: Histograms determined in the absence and presence of 5-HT using transfected, neomycin-resistant cells. **B**: Histograms determined in the same way as in (**A**) using neomycin-resistant cells that had been sorted based on a positive 5-HT response and grown for one to two weeks. The cells have been enriched from 30 to 85% responsive to 5-HT.

detected and the pharmacological profile of the response was determined and compared to that of the 5-HT1c transfected cells (see *Figure 6*).

5.2 Physiology

An advantage of transfecting cell lines to study cell-surface receptors is that it allows the receptor of interest to be studied in an environment devoid of other closely related receptors that may normally be found in the native membrane. Further, it provides a way to study receptor function following carefully specified modifications of the molecular structure of the receptor using techniques such as site-directed mutagenesis. Such structure–function studies require an assay that will be sensitive to subtle changes in function. If

Isolation and characterization of transfected cell lines

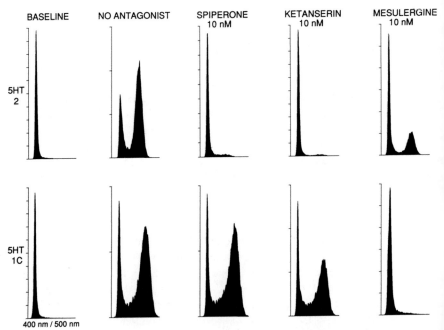

Figure 6. Comparison of antagonist sensitivity in NIH-3T3 cells transfected with cloned 5-HT1c and 5-HT2 receptors. The dual syringe injector system was used with cells plus antagonist in one syringe and 10 μM 5-HT in the other. (Taken from ref. 25.)

possible, it should also provide a means for screening a large population of transfected cells, since expression levels can vary within the populations (see *Figure 5A*). Flow cytometric screening of $[Ca^{2+}]_i$ is well-suited to these requirements.

The standard sample delivery system of the flow cytometer provides a way to study the mechanisms underlying the response to agonist in transfected cells. An example of how this can be accomplished is shown in *Figure 3*. *Figure 3A* shows the time-course of the $[Ca^{2+}]_i$ response to MCh in a human astrocytic cell line. When EGTA was added to the bath in sufficient concentration to drop bath $[Ca^{2+}]$ to less than 10^{-6} M, the late, sustained portion of the response was blocked, suggesting that it was dependent on entry of Ca^{2+} from the extracellular solution. In *Figure 3B*, EGTA is added before the agonist. In this case, the transient early response remains even though the late portion of the response is blocked. This suggests that the early transient response is due to release of Ca^{2+} from intracellular stores. Manipulations of the extracellular $[Ca^{2+}]$ like these can be exploited to determine the nature of a response for new, poorly characterized receptors. It can also be used, in combination with site-directed mutagenesis, to study structure–function relationships.

5.3 Pharmacology

Pharmacological studies and gene cloning have established that multiple receptor subtypes exist for most, if not all, neurotransmitters. The physiological flow cytometric assay can be used to identify and characterize the pharmacological profile of the various members of these receptor families. In addition, cloning by homology can sometimes lead to the identification of 'orphan' receptors for which the natural ligand is unknown. Under these circumstances, flow cytometry can be used to screen cocktails of potential agonists as one approach to receptor type identification. Alternatively, in the case when receptors for a specified ligand are cloned and confirmation of receptor subtype is necessary, an appropriate array of agonists and antagonists can be used to determine the pharmacological profile of the transfected receptor.

For example, the various 5-HT receptor subtypes have characteristic differential affinities for a battery of 5-HT antagonists. The 5-HT1c receptor binds mesulergine with high affinity, mianserin with moderately high affinity, and spiperone and ketanserin with low affinity. The 5-HT2 receptor displays higher affinity than the 5-HT1c receptor for spiperone and ketanserin (26). When the 5-HT1c receptor was cloned and transfected into NIH-3T3 cells, the flow cytometric Ca^{2+} assay was used to establish the pharmacological profile of the cloned receptor. In these experiments, suspensions of indo-1-loaded cells contained one of the antagonists before and during analysis on the flow cytometer. The dual syringe injector system was used, with 5-HT in the other injector. The results (*Figure 7*) are consistent with reported receptor antagonist sensitivity for 5-HT1c receptors, with mesulergine being the most effective at inhibiting increases in $[Ca^{2+}]_i$ in response to 5-HT followed by mianserin. Spiperone was essentially ineffective in blocking response to 5-HT (23).

A direct comparison of 5-HT1c and 5-HT2 receptors in transfected cells using $[Ca^{2+}]_i$ measurements to assay the effectiveness of different 5-HT antagonists in blocking 5-HT-evoked Ca^{2+} mobilization (25) yielded results consistent with those obtained with natural receptors. The differential affinity of the various 5-HT antagonists, ketanserin, spiperone, and mesulergine, for the 5-HT1c and 5-HT2 receptors, are demonstrated in *Figure 6* (25).

6. Conclusion and summary

There are a number of advantages to using flow cytometry to study populations of cells. A flow cytometer is capable of examining each cell of a population, at a rate of up to 5000 cells per second. This makes it possible to detect physiological changes in a small subpopulation of cells in a non-invasive way. Other single-cell analysis methods such as microscopy can be used but are a laborious way to screen for a small subpopulation. Fluorimetry

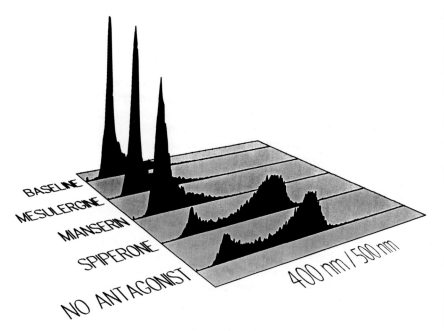

Figure 7. Antagonist sensitivity of the 5-HT1c receptor expressed in transfected NIH-3T3 cells. Baseline data were measured in the absence of 5-HT while all the other histograms were obtained in the presence of 1 μM 5-HT using the dual syringe injector system. All antagonists were used at 1 μM. Antagonists were aded to the syringe with cells. (Taken from ref. 23.)

may be used to screen large populations of cells. However, physiological responses from a small subpopulation may not cause enough change in the total fluorescence of the whole population to be detectable. In flow cytometry the fluorescence of each cell is measured individually and thus is a more sensitive indicator of small subpopulations. Other advantages of flow cytometry include the ability to collect several other parameters besides fluorescence that are useful. Light scatter measurements are valuable for identifying populations of cells based on size and eliminating dead cells, debris, and cell doublets from the collected data. Since flow cytometry data can be displayed as a function of time, the changing distribution of responses of the population of cells can give an indication of the time-course of physiological changes in a population with maintained sensitivity to subpopulation differences. Finally, the cytometer is also a sorter and thus cells can be purified based on their physiological response.

References

1. Kung, P. C., Goldstein, G., Reinherz, E. L., and Schlossman, S. F. (1979). *Science*, **206**, 347.
2. Grynkiewicz, G., Poenie, M., and Tsien, R. Y. (1985). *J. Biol. Chem.*, **260**, 3440.
3. Minta, A., Kao, J. P. Y., and Tsien, R. Y. (1989). *J. Biol. Chem.*, **264**, 8171.
4. Chused, T. M., Wilson, H. A., Greenblatt, D., Ishida, Y., Edison, L. J., Tsien, R. Y., and Finkelman, F. D. (1987). *Cytometry*, **8**, 396.
5. Schieren, I. and MacDermott, A. B. (1988). *J. Neurosci. Meth.*, **26**, 35.
6. MacDermott, A. B., Mayer, M. L., Westbrook, G. L., Smith, S. J., and Barker, J. L. (1986). *Nature*, **321**, 519.
7. Iino, M., Ozawa, S., and Tsuzuki, K. (1990). *J. Physiol. London*, **424**, 151.
8. Hollmann, M., Hartley, M., and Heinemann, S. (1991). *Science*, **252**, 851.
9. Perez-Reyes, E., Kim, H. S., Lacerda, A. E., Horne, W., Wei, X., Rampe, D., *et al.* (1989). *Nature*, **340**, 233.
10. Dohlman, H. G., Caron, M. G., and Lefkowitz, R. J. (1987), *Biochemistry*, **26**, 2657.
11. Berridge, M. J. (1987). *Ann. Rev. Biochem.*, **56**, 159.
12. Roe, M. W., Lemasters, J. J., and Herman, B. (1990). *Cell Calcium*, **11**, 63.
13. Almers, W. and Neher, E. (1985). *FEBS Lett.*, **192**, 13.
14. MacIntyre, E. H., Ponten, J., and Vatter, A. E. (1972). *Acta Pathologica Microbiol. Immunol. Scand., A Pathology*, **80**, 267.
15. Williams, D. A. and Fay, F. S. (1990). *Cell Calcium*, **11**, 75.
16. Kachel, V., Kempski, O., Peters, J., and Schodel, F. (1990). *Cytometry*, **11**, 913.
17. Tsien, R. and Pozzan, T. (1989). *Methods in enzymology*, Vol. 172 p. 230. Academic Press, Orlando, Florida.
18. Miller, D. J. and Moisescu, D. G. (1976). *J. Physiol.*, **259**, 283.
19. Ashley, C. C. and Moisescu, D. G. (1977). *J. Physiol. London*, **270**, 627.
20. Moisescu, D. G. and Thieleczeck, R. (1978). *J. Physiol. London*, **275**, 241.
21. Stephenson, D. G. and Williams, D. A. (1981). *J. Physiol. London*, **317**, 281.
22. Julius, D., MacDermott, A. B., Axel, R., and Jessell, T. M. (1988). *Science*, **241**, 558.
23. Julius, D., MacDermott, A., Jessell, T., Huang, K., Molineaux, S., Schieren, I., and Axel, R. (1988). *Molecular biology of signal transduction. Cold Spring Harbor Symp. Quant. Biol.*, **53**, 385.
24. Kirschmeier, P. T., Housey, G. M., Johnson, M. D., Perkins, A. S., and Weinstein, I. B. (1988). *DNA*, **9**, 219.
25. Julius, D., Huang, K. N., Livelli, T. J., Axel, R., and Jessell, T. M. (1990). *Proc. Natl. Acad. Sci. USA*, **87**, 928.
26. Peroutka, S. J. (1988). *Ann. Rev. Neurosci.*, **11**, 45.

8

Regulation of gene expression in neuronal cell lines

KAREN A. LILLYCROP, CAROLYN L. DENT, and
DAVID S. LATCHMAN

1. Introduction

Ideally, the application of molecular techniques to the study of gene regulation in a particular tissue requires the isolation of large amounts of a single-cell type in which these processes can be analysed and into which DNA constructs derived from regulated genes can be introduced. Unfortunately, virtually none of these criteria are fulfilled in the case of neuronal cells *in vivo*. Thus, neurons from anatomically-defined structures can only be obtained in relatively small amounts whilst brain tissue, although available in relatively large amounts is highly heterogeneous containing many different neuronal and non-neuronal cell types. Moreover, even if neurons are identified, isolated and studied in culture, they are often difficult to transfect with DNA constructs so that the processes regulating the activity of a particular gene promoter cannot be studied in this way.

The availability of immortalized neuronal cell lines overcomes these problems, in that it provides a large-scale source of homogeneous material in which the expression of a particular gene can be studied whilst such cells are frequently relatively easy to transfect allowing the processes regulating a particular gene promoter to be analysed. In this chapter we will therefore discuss the methods which can be used, in conjunction with neuronal cell lines, to study the expression of a particular gene, the activity of its corresponding promoter and the transcription factors which interact with it.

2. Gene expression

The normal starting point for a study of gene regulation in a neuronal cell line would be the finding that a particular protein is present in these cells and absent in another cell type, or that the level of this protein is altered in these cells in response to a specific stimulus such as treatment with nerve growth factor. Clearly, such differences or alterations in the levels of a specific protein can occur by a number of different mechanisms including alterations in the level of gene transcription and alterations in the stability or translatability of the resulting RNA (for review, see ref. 1). In this section we will therefore first consider methods for determining the level of a specific RNA in a particular cell type, thereby determining whether changes in the levels of a particular protein are paralleled by differences in the level of its corresponding RNA. Subsequently, we will discuss the method of measuring directly the transcription rate of a specific gene in particular. Taken together these methods allow the investigator to determine whether changes in the level of a particular protein are brought about by alterations in gene transcription, changes in RNA stability without alterations in transcription or changes in RNA translatability without alteration in RNA level.

2.1 Detection and quantitation of RNA

Many methods exist to detect and quantitate the level of a specific RNA. The choice of method to be used will depend on the number of samples to be processed, the possible existence of different RNAs from the same gene and the sensitivity required to detect a particular RNA. Several alternative methods will therefore be discussed with an indication of the advantages of each.

2.1.1 Dot or slot blotting

A simple rapid method of quantitating the levels of a specific RNA in a number of samples is dot or slot blotting (2). Dot hybridizations were originally performed by spotting a small sample of the RNA preparation onto nitrocellulose and hybridizing to a specific DNA or RNA probe. More recently, filtration manifolds have been designed to accept a large number of samples via circular (dot blotting) or rectangular (slot blotting) slots and to deposit the RNA by vacuum suction on to the nitrocellulose in a fixed pattern that allows the results to be quantitated by scanning densitometry. These manifolds are available commercially (for example, Minifield II, Schleicher & Schuell).

Following spotting of the RNA sample, the filter is hybridized with an appropriate radioactive probe for the gene in question. Another replicate filter is hybridized with a probe for an RNA whose level is likely to be the same in all samples to control for variations in the total amount of RNA in each sample loaded. The choice of probe for this control experiment will vary with the situation, but widely used examples are probes for ribosomal RNA or the actin mRNA which will be expressed in all cell types. Following washing, the filters can be autoradiographed and the amount of the control RNA and the experimental RNA quantitated by densitometric scanning of the autoradiograph, or by cutting out the labelled dots/slots and determining the amount of radioactivity by liquid scintillation counting.

Protocol 1. RNA slot blotting

1. Wet a piece of nitrocellulose briefly in water and soak it in 20 × SSC (3 M NaCl, 300 mM sodium citrate) for 1 h at room temperature (RT). Meanwhile, clean the manifold carefully with 0.1 N NaOH and then rinse it well with sterile water.
2. Place two sheets of 3MM Whatman paper, previously wetted with 20 × SSC, on the top of the vacuum unit of the apparatus. Place the wet nitrocellulose on the bottom of the sample wells cut into the upper section of the manifold. Clamp the two parts of the manifold together, and connect the vacuum unit to a vacuum line.
3. Fill all of the slots with 10 × SSC, and apply gentle suction until all of the fluid has passed through the nitrocellulose filter.
4. Mix the RNA (dissolved in 10 µl of water) with
   - 100% formamide   20 µl
   - formaldehyde   7 µl
   - 20 × SSC   2 µl

   Incubate the mixture for 10 min at 65 °C, and then cool the samples on ice.
5. Add 2 vol. of 20 × SSC to each sample.
6. Load the samples into the slots and then apply gentle suction. After all of the samples have passed through the filter, rinse each of the slots twice with 1 ml of 10 × SSC.
7. After a second rinse has passed through the filter, continue suction for 5 min to dry nitrocellulose filter.
8. Remove the nitrocellulose filter from the manifold, and allow to dry completely at room temperature (RT). Bake the filter for 2 h at 80 °C.
9. Hybridize the filter to a radiolabelled probe as described for Northern hybridizations (*Protocol 3*).

This method thus offers a very rapid means of quantitating the RNA level in a number of different samples. It does not, however, provide any indication of whether the RNA being detected is of the correct size and is of no use where it is desired to quantitate the levels of two different RNAs which hybridize to the same probe.

2.1.2 Northern blotting

In view of the limitations of dot blotting noted above, it is necessary to use alternative methods where it is desired to confirm that the RNA being quantitated is of the correct size. To do this Northern blotting is used, the RNA being first separated by size on an agarose gel and then transferred to a nitrocellulose filter (*Protocol 2*). The filter is then pre-hybridized and subsequently hybridized to an appropriate ^{32}P labelled probe (*Protocol 3*). Following washing and autoradiography, the levels of a particular mRNA can be quantitated approximately by eye or more precisely by densitometric scanning of the autoradiograph. As with dot blotting, another probe for an RNA whose expression is likely to be similar in all samples is hybridized to a replicate filter or subsequently to the same filter to control for any variations in the amounts of total RNA in each sample.

Protocol 2. Northern blotting (adapted from ref. 3)

1. Incubate the RNA samples at 65°C for 10 min in the following solution:
   - RNA (20 µg total RNA or 1 µg Poly A$^+$)   12 µl
   - formamide (deionized)   25 µl
   - 10 × Mops buffer[a]   5 µl
   - formaldehyde   8 µl
2. Chill on ice and add 5 µl 50% (v/v) glycerol containing 0.1 mg/ml bromophenol blue (BPB).
3. Run on a 1–1.5% agarose gel, prepared as follows:
   - 1–1.5 g agarose
   - 10 ml 10 × Mops buffer
   - 75 ml water

   Dissolve the agarose by boiling and cool to 50°C. Add 17 ml formaldehyde (40% w/v solution). Mix and pour immediately.
4. Load the RNA samples on to the gel, use a running buffer of 1 × Mops and electrophorese, until the BPB dye is three-quarters of the way down the gel.
5. To size the RNA transcripts run on the Northern gel, either a commercially available RNA size ladder (Gibco-BRL) or a total RNA sample is used. The marker track on the gel is subsequently stained with ethidium bromide to detect the bands in the RNA ladder or the 28S/18S ribosomal bands in total RNA.[b]
6. To transfer the RNA from the gel on to a hybridization membrane (Hybond N, Amersham) set up a transfer apparatus as follows:

   Set up a platform above a reservoir of 20 × SSC. Cover the platform with 3MM Whatman paper soaked in 20 × SSC (3 M NaCl, 300 mM trisodium citrate). Use a wick of paper, cut to the same width of the gel, but twice as long, so that the ends of the paper dip in the reservoir. The gel is placed on top of the wick and the nitrocellulose membrane placed on top of the gel. A stack of absorbent paper is then placed on top and covered with a glass plate. Capillary transfer is allowed to proceed overnight.

7. Wash the membrane in 2 × SSC, and air dry it.
8. Wrap the membrane in Saran-wrap and irradiate with the RNA side down on a UV transilluminator (254 nm) for 2 min.

[a] 10 × Mops buffer: 0.2 M Mops (3-(N-morpholino) propane–sulphonic acid) 0.05 M Na acetate pH 7.0, 0.01 M EDTA).
[b] To stain a Northern gel, the gel is soaked in 10% TCA, for 10 min, washed in 1 M Tris–HCl pH 8.0 for 10 min, and stained with ethidium bromide (10 mg/ml) for 30 min.

Protocol 3. Northern blot hybridization

1. Place the filter in 30–40 ml of pre-hybridization buffer in a plastic bag and seal it. Pre-hybridization buffer[a] has the following composition:
   - 100 mg/ml denatured herring sperm DNA (Sigma)
   - 5 × SSC
   - 7% SDS
   - 10 × Denhardt's solution[b]
2. Incubate with gentle agitation at 65°C for 2–3 h.
3. Discard the pre-hybridization buffer and replace with hybridization buffer (pre-hybridization solution plus 10% dextran sulfate).
4. Denature the oligolabelled probe and add it to the bag. The probe should be at a specific activity of $>10^8$ c.p.m./mg and at a concentration of 10^6 c.p.m./ml.[c]
5. Re-seal the bag and mix the probe thoroughly with the hybridization solution.
6. Hybridize for 20 h at 65°C with gentle agitation.
7. Wash the filter in three changes of 1 × SSC/0.1% SDS at RT for 15 min followed by two 20-min washes in 0.1 × SSC/0.1% SDS at 65°C.
8. Wrap the filter in plastic film while still damp and expose to X-ray film.
9. To reuse the filter, remove the probe by placing the filter in a tray containing sterile distilled water at 100°C for 5 min.

[a] To dissolve the pre-hybridization solution, warm to 65°C.
[b] Denhardt's solution (100 times) is 2% (w/v) bovine serum albumin, 2% ficoll, 2% polyvinyl pyrollidone. (All molecular biology grade; Sigma.)
[c] The probe can be most conveniently labelled by the random priming method of Feinberg and Vogelstein (4).

Northern blotting thus provides a relatively simple means of quantitating an RNA and is of particular use where multiple RNAs are present since it allows sizing and quantitation of each RNA individually (*Figure 1*).

Regulation of gene expression in neuronal cell lines

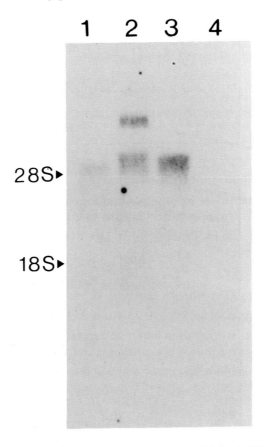

Figure 1. Northern blot hybridization with a radioactively labelled cDNA probe derived from the mRNA encoding the Oct-2 transcription factor. The RNA samples were obtained from whole brain (*track 1*), the neuronal cell line ND7 (*track 2*), B cells (*track 3*) and 3T3 fibroblasts (*track 4*). Arrows indicate the positions of the 28S and 18S ribosomal RNA markers. Note the presence of two distinct mRNAs hybridizing to the probe in the ND7 sample which would not be distinguished by dot blotting. (For further details, see ref. 32.)

2.1.3 RNase protection

Although Northern blotting offers a convenient means for quantitating RNA it may not be sufficiently sensitive to readily detect RNAs of low abundance. Moreover, Northern blotting will detect any RNA which is sufficiently related to the probe to cross-hybridize to it under the conditions used even if the RNA is not identical to the probe. Hence, it does not provide any information on how closely related the RNA being detected is to the probe used. To provide this information and increase sensitivity RNase protection is used (5).

In this technique (*Figure 2*) a cDNA clone derived from the mRNA of interest is cloned into a plasmid vector containing a promoter for transcription

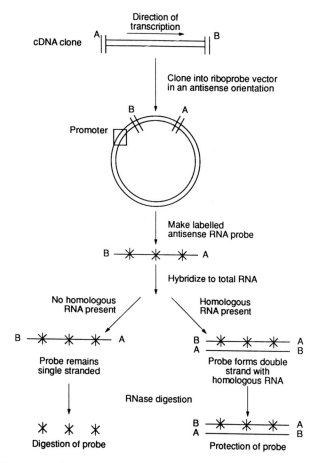

Figure 2. RNase protection assay for the quantitation of specific mRNAs.

by a bacteriophage polymerase such as T7, T3, or Sp6, and a radiolabelled antisense RNA copy is made by transcription with the polymerase *in vitro* (*Protocol 4*). This labelled RNA is then hybridized to the RNA sample being analysed and the mixture digested with ribonucleases A and T1 which digest only single-stranded RNA but do not affect double-stranded RNA (*Protocol 5*). If the sample contains an mRNA homologous to the labelled RNA probe, a double-stranded RNA molecule will form which will not be digested with RNase. Hence, the amount of a particular RNA can be quantitated by measuring the amount of the labelled probe which is protected from digestion (*Figure 2*). Moreover, the ribonucleases will cleave double-stranded RNA at any mismatch between the labelled RNA probe and the RNA being quantitated. Hence, complete protection of the probe will indicate perfect identity of the RNA in the sample with the probe, whereas the presence of smaller

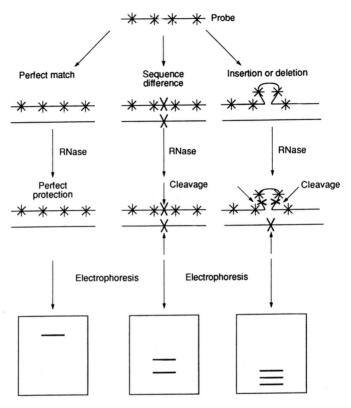

Figure 3. Use of the RNase protection assay to detect mismatches between the probe used and the mRNA which is present in a particular tissue. Note that the assay can detect individual base differences in the two sequences caused, for example, by their being derived from different genes as well as insertions or deletions caused, for example, by alternative splicing.

protected fragments will indicate the position of any mismatches produced by sequence differences or by insertions or deletions in one sequence relative to the other (*Figure 3*).

Protocol 4. RNA transcription *in vitro* to produce a radioactively labelled product

A. *Template preparation*

1. DNA (10–20 μg) containing the insert of interest is linearized with the appropriate restriction enzyme.
2. The cut DNA is then extracted with phenol/chloroform and ethanol precipitated.

3. The template is resuspended at a concentration of 1 µg/µl.

B. *RNA transcription*

1. In a microcentrifuge tube, mix in order:
   - 5 µl   5 × transcription buffer[a]
   - 5 µl   100 mM DTT
   - 0.5 µl RNase inhibitor (25 U/µl. Promega)
   - 5 µl   2.5 mM each of ATP, GTP, UTP, 50 µM CTP[b]
   - 5 µl ^{32}P-CTP (50 µCi at 10 mCi/ml)
   - 2 µl   linearized template (1 µg)
   - 1 µl   SP6 RNA polymerase or T7 RNA polymerase (as appropriate) (at 15–20 U/µl) (Stratagene or Promega)

   Add water to a final volume of 25 µl.

2. Incubate at 37°C for 40 min.

3. To remove the DNA template after transcription add 1 µl RNase-free DNase to a concentration of 1 unit/µg DNA. Incubate for 15 min at 37°C.

4. Make the volume up to 100 µl with water and extract with an equal volume of phenol/chloroform.

5. Ethanol precipitate the RNA by addition of 20 µl of 10 M ammonium acetate and 300 µl absolute ethanol. Stand on dry ice for 10 min and recover RNA by microfuging for 10 min.

[a] 5 × Transcription buffer for Sp6 polymerase 200 mM Tris–HCl pH 7,5, 30 mM MgCl$_2$, 10 mM spermidine. For T3 or T7 polymerases add also 50 mM NaCl.
[b] To produce an unlabelled template, the ^{32}P labelled CTP is omitted and the concentration of unlabelled CTP increased to that of the other three nucleotides.

Protocol 5. RNase protection

1. Single-stranded radiolabelled RNA transcripts are synthesized. DNase treated, extracted with phenol/chloroform and ethanol precipitated as described in *Protocol 4*.

2. For RNase protection the RNA transcripts are then purified on an acrylamide sequencing type gel. The RNA pellet is redissolved in a small volume (20 µl) of formamide dye mix (80% formamide, 5 mM EDTA pH 8.0, 0.2% bromophenol blue, 0.2% xylene cyanol, heated to 90°C for 5 min and loaded directly on to a 6% acrylamide/7 M urea gel. Electrophorese for 2–3 h at 30 W.

3. After separating the full-length riboprobe the gel is autoradiographed. To do this, separate the sequencing plates, place hot ink markers around

Protocol 5. *Continued*

    the gel, cover with Saran-wrap and expose to film in an X-ray cassette for 10 min.

4. Using the exposed film as a template, the full-length transcripts are excised from the gel.
5. The riboprobe is eluted by shaking in a 37°C rotating shaker overnight in 400 μl of elution buffer (0.5 mM ammonium acetate, 10 mM MgAc, 1 mM EDTA, 0.1% SDS). The supernatant is transferred to a fresh microcentrifuge tube, 10 μg of carrier tRNA and 1 ml of ethanol are added. Stand on dry ice for 10 min and recover the RNA by microfuging for 10 min.
6. After drying down the pellet, the probe is dissolved in 500 μl of water and 2 μl is counted to assess incorporation of ^{32}P.
7. 2 to 7×10^5 c.p.m. of gel purified riboprobe is added to 5 to 20 μg of total cellular RNA or 2 μg of poly A and co-precipitated with ethanol.
8. The pellets are redissolved in 30 μl of hybridization buffer (80% formamide, 40 mM Pipes, 400 mM NaCl, 1 mM EDTA), heated to 85°C for 20 min and then incubated O/N at 56°C.[a,b]
9. To each hybridization, 300 μl of RNase digestion buffer is added (10 mM Tris–HCl, pH 7.5, 5 mM EDTA, 300 mM NaCl, 20 μg/ml RNase A, 1 μg/ml RNase T1). Incubate at 30°C for 30 min.
10. Add 20 μl of 10% SDS and 50 μg/ml proteinase K, and incubate at 37°C for 15 min.
11. Extract samples with an equal volume of phenol/chloroform.
12. To the aqueous phase add 5 μg of carrier tRNA and ethanol precipitate in dry ice for 10 min. Microfuge for 10 min, wash pellets in 70% ethanol, and dry under vacuum.
13. Dissolve pellets in 2 μl of formamide dye. Heat to 90°C for 5 min and load on to a 6% acrylamide/urea gel, alongside molecular weight markers. Electrophorese for 3 h at 30 W.
14. Perform autoradiography of the gel using an intensifying screen.

[a] Tubes are microfuged after each step and sealed with parafilm for the overnight incubation.
[b] The hybridization temperature may need to be changed depending on the G/C content of the sequence.

RNase protection can therefore be used to quantitate the levels of RNA in a particular sample. Moreover, since cleavage will occur at the site of any mismatch it will be possible to determine whether the RNA in the sample is identical to the probe used, and to map the sites of any mismatches by determining the sizes of any fragments which form (*Figure 3*). This will allow the identification of sequence differences caused by the probe and the sample RNA being derived from different genes, as well as the position of any insertions or deletions in one sequence, compared to the other which may arise by variations in the splicing pattern of a particular RNA in different tissues (for review, see ref. 6).

2.1.4 Quantitation of mRNA by PCR

Some RNAs (for example, those encoding some transcription factors) will be present at such low levels that even RNase protection may not be sufficiently sensitive for their detection and quantitation. In these cases, very sensitive detection of the RNA can be achieved by using the polymerase chain reaction (PCR; ref. 7) in conjunction with specific primers to amplify the RNA of interest. Because PCR will only amplify DNA and not RNA it is necessary to first prepare a cDNA copy of all the mRNAs in the cell, using reverse transcriptase (*Protocol 6*), and specific primers derived from the sequence of the mRNA of interest can then be used to specifically amplify the cDNA/mRNA hybrid produced in this manner (*Protocol 7*).

Protocol 6. cDNA synthesis

1. Assemble the following reagents in a final volume of 20 μl: 1 × PCR buffer,[a] 1 mM each dNTP, 1 unit/μl of RNase inhibitor, 100 pmol of random hexamer oligonucleotides (Pharmacia), 1–5 μl of RNA sample (0.2–1 μg of total RNA is sufficient)[b] and 200 units of M.MuLV reverse transcriptase.
2. Incubate the reaction at 37°C for 1 h.
3. Verify that no DNA is present in the RNA samples by running control samples without reverse transcriptase, prior to PCR (*Protocol 7*).

[a] 10 × PCR buffer is 200 mM Tris–HCl pH 8.4, 500 mM KCl, 25 mM $MgCl_2$, 1 mg/ml bovine serum albumin.
[b] It is helpful to heat the RNA to 65°C for 3 min, then quick-chill it on ice before addition of the mRNA to the reaction mixture, to break up aggregates and secondary structure.

Protocol 7. PCR reaction

1. To the 20 μl of reverse transcriptase reaction add 80 μl of 1 × PCR buffer containing 25 pmol each of upstream and downstream primer and 1 to 2 units of *Taq* polymerase.[a]
2. To prevent evaporation during PCR, layer 100 μl of mineral oil on top of the PCR mixture.
3. The number of PCR cycles required depends on the abundance of the target. Usually, somewhere between 20 to 50 cycles is required, but the

Protocol 7. *Continued*

optimal number to give a signal which is linearly related to mRNA level should be determined in each case by amplifying a dilution series of an RNA sample using different numbers of cycles.

4. A PCR profile that works well for amplification of a target less than 500 bp is denaturing for 30 sec at 94°C, annealing for 30 sec at 2°C below the T_m^b of the primers, and extending for 30 sec at 72°C.

5. After amplification, the mineral oil is removed by extraction with 100 µl of TE saturated chloroform. The upper aqueous layer is saved and 20 µl analysed on a 2% agarose gel.

[a] If the primers are located further than 600 bp apart, 1 mM of each dNTP is added to the PCR reaction mixture.
[b] To calculate the T_m of the PCR primers (less than 50 bp) the following equation can be used: $T_m = 4 (G + C) + 2 (A + T)$.

The method described above is excellent for determining whether a particular mRNA is present in a specific cell type. Moreover, as with Northern blotting and RNase protection, the levels of the RNA in different samples can be compared by carrying out a control amplification with primers specific for an RNA whose level should be identical in the different samples. Comparison of the signal obtained in each sample with the primers specific for the control or the experimental mRNA will give a measure of the relative level of the experimental RNA in each sample. Occasionally, however, differences in amplification efficiency of a particular RNA can occur between samples using this method resulting in artefactual apparent differences in RNA level. To control for this, it is best to use an internal control template in each sample which is co-amplified with the RNA under test. Although this can be achieved by amplifying the control RNA with its specific primers in the same tube as the test RNA, it is preferable to use a control template which will be amplified with the same primers as the experimental RNA but whose PCR product can be distinguished from the product of the RNA following amplification (*Figure 4*). By this means, the procedure is directly controlled for differences in amplification by the test primers.

This is achieved by using a control template which can be amplified by the same primers as the experimental RNA but which differs from it by the presence of a restriction enzyme site allowing its PCR product to be distinguished from that of the test RNA. Normally, this template would be prepared by using a cDNA clone of the test RNA which has been modified to contain a restriction site by site-directed mutagenesis (8) or by using a cDNA clone of the homologous RNA from a related species which differs sufficiently in sequence to result in differences in its sites for restriction enzymes but which retains the sequences recognized by the PCR primers. This cDNA clone is transcribed into an unlabelled RNA using T7 or Sp6 polymerase *in vitro* (*Protocol 4*, note *b*) and the resulting RNA added to each

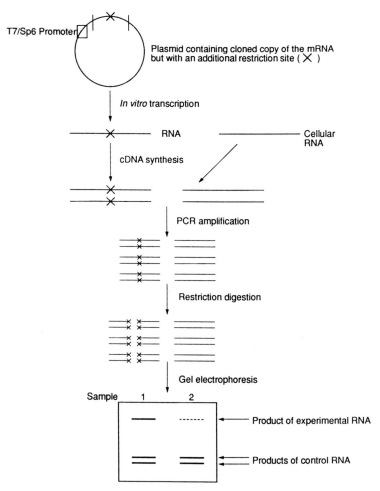

Figure 4. Use of the polymerase chain reaction (PCR) to quantitate specific mRNAs. The mRNA is quantitated against a co-amplified control RNA whose amplification product can be distinguished from that of the mRNA by restriction enzyme digestion.

PCR reaction (*Protocol 8*). Following the PCR reaction, the products are digested with the appropriate restriction enzyme and the fragments produced by the control template and the experimental RNA separated by agarose gel electrophoresis (*Figure 4*).

Protocol 8. Quantitative PCR for mRNA

The procedure is carried out as in *Protocol 7* with the following variations:

1. Master mixes containing all the reaction components for both the cDNA (minus the RNA) and PCR (minus the cDNA) reaction are prepared.

Protocol 8. *Continued*

2. An aliquot of the *in vitro* synthesized RNA, is added to the cDNA master mix.
3. The number of PCR cycles performed is crucial, if the assay is to be quantitative. If too many cycles are performed, the amount of PCR product may exceed the linear range of the detection methods (for example, ethidium bromide gel staining and/or Southern blotting) and therefore differences between mRNA levels will be difficult to determine.

This method therefore provides a precise means of quantitating RNAs which are present at extremely low levels in the cell line being studied.

2.2 Measurement of transcription rates

The methods described in the previous section will allow investigation of whether differences in protein levels in different cell types or following particular treatments are paralleled by differences in the level of the corresponding mRNA. If this is the case, such differences could be dependent on differences in the transcription rate of the specific gene encoding the mRNA or on post-transcriptional changes such as alterations in mRNA stability. To distinguish these possibilities it is necessary to directly measure gene transcription using a nuclear run on assay (*Figure 5*). In this method (9, 10) nuclei are isolated free of cytoplasm (*Protocol 9*). This results in the nuclei being deprived of the cytoplasmic source of ribonucleotides and transcription ceases, with the RNA polymerase molecules stalled at the point within the gene they had reached when the cells were lysed. If labelled ribonucleotide is now added no new initiation of transcription will occur but the stalled RNA polymerase molecules will 'run-on' to the end of the gene incorporating label into the corresponding RNA (*Protocol 10*). Obviously, the more RNA polymerase molecules which are stalled on the gene, the greater the amount of label incorporated into a particular RNA. This can be detected by using the labelled RNA products of the run-on reaction to probe a dot blot filter on to which plasmid DNAs containing the gene of interest have been spotted (*Protocol 11*). As the initial binding of RNA polymerase to the gene is the rate limiting step of transcription, the amount of label binding a particular cloned gene on the dot blot filter is a measure of gene transcription.

Protocol 9. Preparation of nuclei

Carry out all procedures at 4°C.

1. Lyse cells in 10 mM Hepes pH 7.9, 10 mM NaCl, 3 mM $MgCl_2$, 0.05% NP40.
2. Check cell lysis is complete and nuclei are intact by microscopic examination.

3. Spin nuclei through lysis buffer containing 30% sucrose at 4000 g for 3 min in a microfuge (low-speed setting).
4. Wash nuclear pellet with lysis buffer and re-spin through lysis buffer containing 30% sucrose.
5. Resuspend nuclei at 2.5×10^7 to 10^8 nuclei per millilitre in 100 µl of 50 mM Hepes pH 7.9, 40% glycerol, 5 mM $MgCl_2$ 0.1 mM EDTA.
6. Quick-freeze and store at −70°C until required.

Protocol 10. Nuclear run-on

1. Make up the following reaction mix:
   - 500 mM Hepes pH 7.9 — 20 µl
   - 1 M KCl — 24 µl
   - 1 M DTT — 4 µl
   - β-mercaptoethanol — 0.4 µl
   - 500 mM magnesium acetate — 2 µl
   - 250 mM manganese chloride — 0.8 µl
   - 500 mM EDTA — 0.4 µl
   - 1 M phosphoenolpyruvate — 1.6 µl
   - 1 M fructose phosphate — 0.4 µl
   - pyruvate kinase (200 units per millilitre) — 2 µl
   - Tween 80 — 2 µl
   - 1 M ATP, CTP, UTP each (Pharmacia) — 2 µl
   - 10 mM GTP (Pharmacia) — 2 µl
   - ^{32}P GTP (10 mCi/ml) (Amersham, NEN) — 5 µl
   - RNase inhibitor (2.5 units/µl) — 2 µl
   - water — 121 µl
   - Total — 200 µl

2. Add 100-µl aliquots of this mixture to a freshly thawed 100-µl sample of nuclei and allow run-on transcription to proceed at RT for 1 h. If necessary alpha-amanitin can be added to the transcription reaction at 2.5 µg/ml to inhibit RNA polymerase II and allow the study of RNA polymerases I and III alone.
   Following completion of the run-on reaction, the labelled RNA products are isolated.

Protocol 10. *Continued*

3. To the run-on reaction add:
   - vanadyl ribonucleoside complex (200 mM; Gibco-BRL)   2 μl
   - transfer RNA (100 mg/ml)   0.8 μl
   - RNase-free DNase I (6 μg/μl)   1 μl

   and digest the DNA in the sample at 30°C for 1 h.

4. Then add:
   - 1 M Tris–HCl pH 7.5   2 μl
   - 500 mM EDTA   2 μl
   - 10% SDS   20 μl
   - 20 mg proteinase K   2 μl

   and digest the DNase at 37°C for 1 h.

5. Extract the RNA with phenol/chloroform.
6. Precipitate the RNA with two volumes of ethanol overnight at −20°C.
7. Spin the labelled RNA pellet and resuspend in 10 mM Tris–HCl pH 7.5, 1 mM EDTA. The amount of labelled RNA is determined by scintillation counting.

Protocol 11. Dot blotting and hybridization with labelled RNA

1. Boil 5–10 μg of plasmid DNA in 0.3 M NaOH for 10 min.
2. Chill the sample on ice.
3. Add ammonium acetate pH 5.5 to 2 M.
4. Apply samples to a nitrocellulose filter as in *Protocol 1*.
5. Rinse the filter in 2 × SSC and bake it for 2 h at 80°C.
6. Pre-hybridize the filter overnight at 42°C in 5–10 ml of the following mixture (for 50 ml):
   - 20 × SSC   10 ml
   - 1 M sodium phosphate   2.5 ml
   - 100 × Denhardt's solution   0.5 ml
   - 20% SDS   0.5 ml
   - 50% formamide   25 ml
   - 100 mg/ml tRNA   0.125 ml
   - water   11.5 ml
7. Add the labelled RNA and incubate for 72 h at 42°C.
8. Wash the filters in 2 × SSC, 0.1% SDS at 65°C for 1 h.
9. Rinse twice in 2 × SSC to remove the SDS.

10. Wash in 2 × SSC containing 20 μg/ml RNase A for 1 h at RT.
11. Wash in 2 × SSC, 0.1% SDS for 1 h at 65°C.
12. As in dot/slot blotting of RNA (*Protocol 1*) the amount of label binding to each DNA can be quantitated by autoradiography followed by densitometric scanning or by cutting out the dots and counting.

In this manner the relative rate of transcription of a particular gene in different cells or different situations can readily be quantitated.

3. Promoter activity

Once it has been shown that the expression of a particular gene is regulated at the level of transcription, it is necessary to map the elements within the gene responsible for this effect. To do this, genomic clones containing the gene in question are isolated and putative regulatory sequences tested for their effect on transcription by linking them to a gene encoding a readily assayable product such as chloramphenicol acetyl transferase (11), beta-galactosidase (12), or luciferase (13). A number of plasmid vectors containing the coding regions of these genes are now available and contain multiple cloning sites upstream of the coding region to facilitate insertion of a heterologous promoter (14). Once this has been done the hybrid construct is introduced by transfection into different cell types or into the same cell type treated in different ways, and any effect of the regulatory sequences on production of the assayable product is assessed. It should be noted that although initially the gene promoter region upstream of the start site of transcription will be assayed in this way, it may be necessary to include elements far upstream of the start site of transcription or downstream of it in order to identify all the regulatory elements of the gene, since these regions often contain enhancer elements which regulate promoter activity (15).

3.1 Transfection

In order to test promoter activity, it is necessary to introduce the construct containing it into cultured cells. A number of techniques exist for doing this, including treatment with calcium phosphate (11), DEAE dextran (16), and electroporation (17). (See also Chapters 2 and 4.) We have found the calcium phosphate procedure to be effective for several different neuronal cell lines and it is therefore presented here (*Protocol 12*).

3.2 Assay of promoter activity

Once the transfection protocol has been carried out, the cells can be harvested and promoter activity determined by assaying the activity of the enzyme encoded by the test gene. The activity of the enzyme following transfection of different cell types or in differently treated cells provides a measure

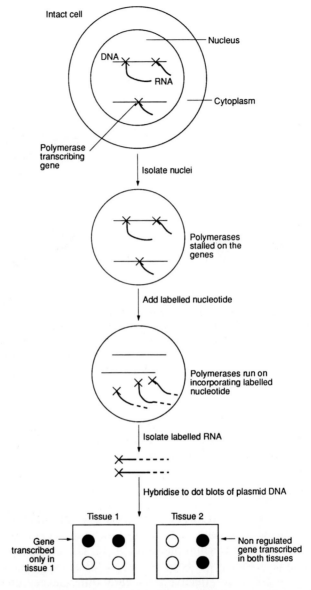

Figure 5. Nuclear run-on assay in which the transcription rate of a specific gene is quantitated by measuring the amount of labelled RNA produced from it by run-on transcription in isolated nuclei. Solid lines indicate unlabelled RNA made prior to isolation of nuclei whilst broken lines indicate radioactively labelled RNA made during run-on transcription. Solid circles indicate radioactive RNA binding to the plasmid DNA; open circles indicate no radioactive binding.

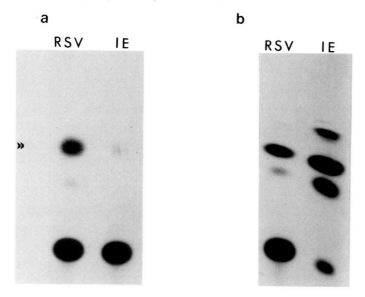

Figure 6. Assay of chloramphenicol acetyl transferase (CAT) activity following transfection of the ND neuronal cell line (**a**) or the BHK-21 epithelial cell line (**b**) with constructs in which expression of the CAT gene is driven by either the Rous Sarcoma virus (RSV) or Herpes simplex virus immediate-early (IE) promoters. Promoter activity is measured by the proportion of chloramphenicol converted into the acetylated form (*arrowed*). Note the very weak activity of the IE promoter in the neuronal cell line compared to epithelial cells. (For further details, see ref. 33.)

of the relative promoter activity under these conditions. In experiments of this sort, however, it is necessary to control for differences in the efficiency of DNA uptake between different cell types or under different conditions. This can be achieved by transfecting with constructs containing another promoter whose activity is unchanged in the different cell types. The constructs containing this promoter are transfected in parallel with those containing the regulated promoter and the activity in the different samples compared (*Figure 6*). However, it is preferable to transfect each cell sample with both the regulated and control promoter constructs. Hence the activity of each promoter can be assessed in the same sample, controlling for variations in transfection efficiency between different plates of cells. Evidently, to do this, the control and regulated promoters must drive the expression of different assayable proteins. We therefore give protocols for assaying the activity of chloramphenicol acetyl transferase (CAT; *Protocol 13*) and beta-galactosidase (*Protocol 14*) in the same extract to allow this to be done. All assays are carried out on samples which have been equalized for their content of total protein as described in *Protocol 15*. The choice of which enzyme should be expressed from the control promoter and which from the regulated one is entirely arbitrary and will depend on the availability of control promoter constructs, vectors, etc.

Protocol 12. The calcium phosphate method

1. On the day before transfection (day 1) replate the cells to be used at a density of $10^4/cm^2$.
2. On day 2, replace the culture medium with 5 ml of fresh medium containing 10% fetal calf serum. DNA is added to the cells 2 h later.
3. To prepare the calcium phosphate–DNA precipitate for a 90-mm dish containing 5 ml of medium, set up the following solutions. In tube A, place a solution containing 5–20 ng of DNA[a] together with 31 ml 2 M $CaCl_2$ and bring the final volume to 0.25 ml with water. To tube B, add 0.25 ml of 2 × HBS.[b]
4. To make the precipitate, the contents of tube A must be added to the HBS in tube B. The order of addition is crucial. Add the DNA solution dropwise to the HBS. The precipitate[c] will form immediately.
5. Pipette the precipitate on to the cells by slightly tilting the dish and adding the precipitate to the medium. Put the cells back into the incubator immediately to ensure that the pH does not change.
6. Incubate the cells for 4 h to 12 h. The longer incubation is sometimes required for promoters that are expressed weakly in neuronal cells.
7. Wash the cells in serum-free medium and then feed them with complete medium.
8. Harvest the cells on day 4. A test of transient expression can be carried out at this stage.

[a] The amount of DNA added to each transfection, should be the same. This can be achieved by adding appropriate amounts of salmon sperm DNA.
[b] 10 × Hepes-buffered saline (HBS). The 10 × stock solution contains 8.18% NaCl (w/v), 5.94% Hepes (w/v), 0.2% Na_2HPO_4 (w/v). The solution is stored at 4°C in 50-ml aliquots. For transfections, use the 10 × solution to prepare a 2 × HBS solution, and adjust the pH to 7.12 with 1 N NaOH. Great care is needed in making up this buffer since the pH is very critical for these experiments.
[c] If the precipitate looks dense and opaque, rather than translucent, the HBS has not been prepared at the correct pH.

Protocol 13. Chloramphenicol acetyl transferase assay

This assay relies on allowing the enzyme to acetylate [^{14}C]-chloramphenicol and assaying the level of acetylated chloramphenicol by thin-layer chromatography (TLC).

1. Following transfection wash the cells with PBS,[a] harvest them and transfer them to a 1.5-ml microcentrifuge tube.
2. Add 100 μl of 0.25 M Tris–HCl pH 7.5 to the cell pellet.

3. Disrupt the cells by freezing and thawing. To freeze–thaw, immerse the tubes in liquid nitrogen for 2 min and then transfer them to a 37 °C water bath. Repeat the cycle three times.
4. Spin down the cell debris and save the supernatant to test for enzyme activity. Samples may be saved at this point by storage at −20 °C.
5. Depending on the cell type and promoter to be assayed, the amount of extract assayed may vary

    The reaction mixture contains:
    - 70 µl   0.25 M Tris–HCl, pH 7.5
    - 35 µl   water
    - 20 µl   cell extract
    - 1 µl   [^{14}C]-chloramphenicol (40–50 Ci/mmol) (Amersham)
    - 20 µl   4 mM acetyl CoA[b]
6. Incubate the reaction mixture for 30 min at 37 °C. The incubation time can be increased up to 60 min provided enough active acetyl-CoA is added to keep the assay linear.
7. Extract the chloramphenicol with 1 ml ethyl acetate, by vortexing for 30 sec.
8. Spin for 2 min in a microcentrifuge tube and save the top organic layer which will contain all forms of chloramphenicol.
9. Dry down the ethyl acetate under vacuum. This will take about 2 h.
10. Resuspend the chloramphenicol samples in 15 µl ethyl acetate and spot them on to silica gel TLC plates.
11. These plates are subjected to ascending chromatography with a 95:5 mixture of chloroform:methanol.[c]
12. After air drying, expose the chromatography plate to X-ray film. After exposure, the regions corresponding to acetylated and non-acetylated chloramphenicol can be cut out and counted.
13. The percentage of total chloramphenicol which has been converted to the monoacetate form gives an estimate of the transcriptional activity.

[a] To prepare 1 litre of PBS dissolve 8 g NaCl, 2 g KCl, 1.5 g Na$_2$HPO$_4$ and 2 g KH$_2$PO$_4$ in water and make up to 1 litre.
[b] 4 mM acetyl-CoA is made up by dissolving 1.5 mg in 0.5 ml water. It is very unstable and should be made up fresh or kept at − 20 °C for not more than 10 days.
[c] The TLC tank should be lined with filter paper around the inside to assist equilibrium. The solvent should be made up fresh every day, since chloroform is very volatile.

Protocol 14. Beta-galactosidase assay (12)

1. Following transfections treat the cells as described in steps 1 to 4 of the chloramphenicol acetyl transferase assay (*Protocol 13*).

Protocol 14. Continued

2. Take 40 µl of cell extract, add 2 µl 0.1 M DTT, 6 µl glycerol and 500 µl of 60 mM Na_2HPO_4, 40 mM NaH_2PO_4 2 H_2O, 1 mM $MgSO_4$ 7 H_2O, 10 mM KCl, 40 mM β-mercaptoethanol. Vortex and incubate for 5 min.
3. Add 100 µl (2 mg/ml) ONPG (O-nitrophenyl-β-D-galactopyranoside), and incubate the reaction at 37°C until a visible yellow colour is achieved. (Can take from 5 min to 24 h.)
4. Stop the reaction with 250 µl 1 M Na_2CO_3.
5. Measure the colorometric change in a spectrophotometer at 420 nm.

Note: It is important to remember that mammalian cells contain a eukaryotic isozyme for beta-galactosidase. Therefore a blank containing a non-transfected cellular lysate should be included in the experiment.

Protocol 15. Bradford assay for protein concentration (18)

1. Following transfection the cells are harvested and then lysed as described in *Protocol 13*. The lysed cells are then spun to remove the cell debris and an estimate of protein concentration can be performed on the supernatant to allow the assay of chloramphenicol acetyl transferase or beta-galactosidase to be carried out on equal amounts of total protein.
2. Add 1 ml of dye reagent[a] to 50 µl of each sample.
3. Measure the absorbance of the sample at 595 nm after 15 min.
4. If the absolute concentration of protein is needed, then a standard curve can be constructed using BSA as standard (draw A_{595} vs. [BSA] mg/ml).

[a] Dye reagent:
- Coomassie brilliant blue G   100 mg
- SDS   30 mg
- 95% (v/v) ethanol   50 ml
- 85% (v/v) phosphoric acid   100 ml

Dilute the mixture to a final volume of 1 litre, using distilled water.

Once a region of the promoter which can confer a pattern of regulation on another gene has been identified, the promoter can be truncated until the effect is lost allowing identification of the precise region of the promoter which confers this effect (see, for example, ref. 19). Subsequently, this region can be cloned into a vector in which a heterologous promoter drives the CAT gene in order to determine if it can confer a specific pattern of regulation on a heterologous promoter (20).

Karen A. Lillycrop, Carolyn L. Dent, and David S. Latchman

4. Identification of DNA-binding transcription factors

The ability of a particular region of the gene promoter to produce a specific pattern of gene expression is normally dependent on its ability to bind one or more specific transcription factors which are present only in a specific cell type or are activated in response to a particular stimulus (for review, see ref. 21). Hence, once a region of the promoter which produces a particular pattern of regulation has been identified, it is necessary to identify the transcription factors which bind to it so they can be characterized and their activity in different cell types and under different conditions investigated. To do this whole-cell or nuclear extracts containing these factors are prepared and their binding to the specific DNA sequences investigated by several different techniques.

4.1 Preparation of cellular extracts

Extracts can be prepared either from whole cells or from isolated nuclei. All procedures are carried out at 4°C and the extracts stored at −70°C after preparation.

Protocol 16. Preparation of nuclear extracts (22)

1. Harvest 5×10^7 to 10^8 cells and wash them with PBS.
2. Resuspend the cells in 5 vol. of hypotonic buffer A (10 mM Hepes pH 7.9, 1.5 mM $MgCl_2$ 10 mM KCl, 0.5 mM DTT) and protease inhibitors, 0.5 mM PMSF. 1 μg/ml leupeptin, 1 μg/ml pepstatin A, 1 μg/ml aprotinin, and 10 mM beta-glycerophosphate) and stand them on ice for 10 min.
3. Spin the cells at 1000 g for 10 min and resuspend in 3 vol. of buffer A.
4. Add NP40 to 0.05% and homogenize the cells with twenty strokes in a tight-fitting homogenizer.
5. Check the samples for release of nuclei by phase-contrast microscopy.
6. Spin at 1000 g for 10 min to pellet the nuclei.
7. Resuspend the nuclei in 1 ml buffer C (20 mM Hepes pH 7.9, 25% glycerol, 1.5 mM $MgCl_2$, 0.25 mM EDTA, with protease inhibitors as in buffer A).
8. Measure the volume of solution and add NaCl to 400 mM. Incubate the solution on ice for 30 min.
9. Spin at 16 000 g for 20 min at 4°C in a refrigerated microfuge.
10. Aliquot the supernatant and snap-freeze in liquid nitrogen. Store aliquots at −70°C.

Protocol 17. Preparation of whole-cell extracts (23)

1. Resuspend the cells in buffer C^a (see *Protocol 16*).
2. Homogenize the cells with twenty strokes in a tight-fitting Dounce homogenizer.^{b}
3. Add NaCl to a final volume of 400 mM and incubate the solution on ice for 30 min.
4. Spin at 16 000 g for 20 min at 4 °C in a refrigerated microfuge.
5. Snap-freeze supernatant aliquots and store them at −70 °C.

^{a} To prepare extracts from tissues, grind the tissue in liquid nitrogen to a fine powder before resuspending in buffer C.
^{b} For tissue extracts it may be necessary to use a tissue macerator instead of a Dounce homogenizer to get efficient disruption of cells.

4.2 DNA mobility shift assay

Once extracts have been prepared from the cells of interest, they can be used in a number of ways to study the proteins binding to a particular sequence. The simplest of these is the gel retardation or DNA mobility shift assay (24, 25) which relies on the principle that a DNA fragment to which a protein has bound will move more slowly in gel electrophoresis than the same fragment without bound protein. Hence, the proteins binding to a particular DNA fragment can be investigated by radioactively labelling the fragments (*Protocol 18*), incubating it with cell extract and electrophoresing on a non-denaturing gel (*Protocol 19*). The retarded DNA-protein complexes can then be visualized by autoradiography (*Figure 7*). Moreover, by carrying out the assay using proteins prepared from different cell types or from the same cell type under different conditions, the nature of the factors binding to a specific piece of DNA in different situations can be determined (*Figure 8*).

Once the proteins binding to this sequence have been identified by this means, it is possible to investigate the precise sequence specificity of this binding. This is achieved by including a large excess of an unlabelled oligonucleotide of specific sequence in the binding reaction. If the protein binding to the labelled oligonucleotide can also bind to this unlabelled oligonucleotide it will do so and the retarded band will disappear (*Protocol 19*, note *b*; *Figures 8* and *9*). The sequence specificity of a novel binding protein which can be determined in this way, may provide clues about its relationship to previously characterized transcription factors with identical or related DNA-binding specificities. Further information about the relationship of a novel factor to known factors can also be obtained by including antibody to a previously characterized factor in the binding reaction. If this antibody reacts with the

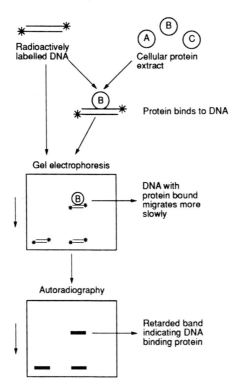

Figure 7. DNA mobility shift assay in which the binding of a protein (B) to a radioactively labelled DNA sequence is detected by its ability to form a slow moving complex with the DNA.

protein of interest it will either prevent its binding to DNA, so abolishing the complex, or produce a so-called supershift of the complex by binding to the DNA-bound protein and decreasing the mobility of the complex (*Protocol 19*, note c; *Figure 10*).

Protocol 18. Labelling of oligonucleotide probes for DNA mobility shift assay.

1. Anneal the separately synthesized strands of the oligonucleotide by heating equimolar amounts of each to 80°C for 2 min then cooling slowly to RT.[a,b]

2. Incubate 2 pmol of annealed oligonucleotide with 20 μCi gamma ^{32}P ATP in 50 mM Tris–HCl pH 7.6 10 mM $MgCl_2$, 5 mM DTT, 0.1-mM EDTA and 4 units of T4 kinase at 37°C for 30 min.

Protocol 18. *Continued*

3. Separate the labelled oligonucleotide from the free probe on a Sephadex G25 column and recover the void volume in 200 μl of STE (10 mM Tris–HCl pH 7.6, 1 mM EDTA, 100 mM NaCl). 1 μl of probe should be sufficient for each band shift incubation.

[a] For fragment probes, the phosphates at the end of the purified fragment must be removed with calf intestinal phosphatase prior to kinase labelling.
[b] For preparing concatamerized oligonucleotide probes for use in South-western blotting (*Protocol 22*) kinase treat the annealed oligonucleotide in 50 mM Tris–HCl pH 7.6, 10 mM $MgCl_2$, 5 mM DTT, then add DNA ligase and ATP to 5 mM and allow to ligate overnight at RT. The kinased, ligated concatamers are then separated from free nucleotide on a Sephadex G50 column by collecting the first peak.

Protocol 19. DNA mobility shift assay

1. Set up a 20 μl binding reaction containing 4% Ficoll, 20 mM Hepes pH 7.9, 1 mM $MgCl_2$, 0.5 mM DTT, 50 mM KCl, 2 μg poly dIdC (Pharmacia), 10 fmol double-stranded end-labelled oligonucleotide or DNA fragment probe and approximately 2 μg of whole-cell or nuclear protein extract.[a,b,c]
2. Incubate on ice for 40 min.
3. Load on to 4% polyacrylamide:bisacrylamide (29:1) gel[d] in 0.25 × TBE[e] and run in 0.25 × TBE at 150 V for approx. 2½ h.
4. Dry the gel[f] under vacuum on to 3MM paper (Whatmann) and autoradiograph.

[a] Prior to use, the protein concentration of different extracts is equalized based on assays of their protein content (*Protocol 15*).
[b] For competitor assay, competitor oligonucleotides are added to the mixture at one-fold, tenfold, and 100-fold molar excess before addition of the extract.
[c] For antibody assay, 1 μl of each of a series of dilutions of the antiserum under test is added to the binding reaction before addition of the extract. Similar dilutions of pre-immune serum are added to parallel reactions as a control.
[d] The gel is pre-electrophoresed before addition of the samples until the current drops from 30 mA to 10 mA (approx. 2 h).
[e] 1 × TBE is 10 mM Tris–HCl, 10 mM boric acid, 2 mM EDTA pH 8.3.
[f] The gel is run until bromophenol blue in a separate marker track has run approximately two-thirds of the way down the gel.

The DNA mobility shift assay thus provides a method of identifying a particular factor which binds to a specific DNA sequence and characterizing its distribution and DNA-binding specificity.

4.3 DNase I footprinting assay

Having identified a protein binding to a specific DNA sequence, the area of contact between the DNA and protein can be localized by using the same

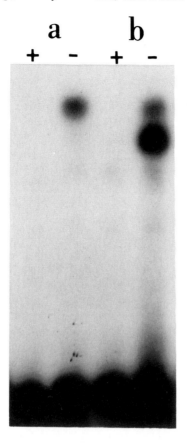

Figure 8. DNA mobility shift assay using extracts prepared from 3T3 fibroblasts (**a**) or the ND neuronal cell line (**b**) and a radioactively labelled octamer DNA sequence (ATGCAAAT). The assays were carried out in the presence (+) or absence (−) of a 100-fold excess of unlabelled competitor octamer DNA sequence. Note that whilst 3T3 cells contain only one protein capable of binding to the octamer sequence, the ND cells contain both this protein and an additional octamer binding protein. (For further details, see ref. 33.)

extracts to carry out a DNase I footprinting assay (26, 27). To do this, a double-stranded DNA fragment labelled at only one end is incubated with the protein extract and then digested with a small amount of DNAse I which will digest the DNA. At the correct level of DNase I each molecule will be cut only once or a very few times by the enzyme, giving rise to a ladder of bands when the sample is run on a denaturing gel. Regions where protein has bound to the DNA, however, will be protected from digestion and hence will appear as a blank area or footprint on the gel (*Figure 11*).

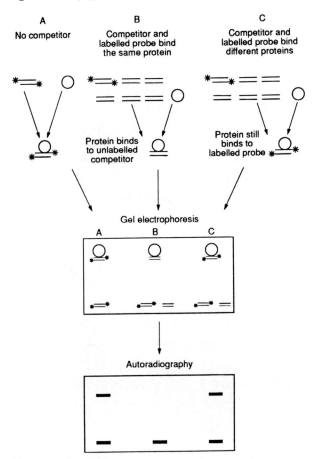

Figure 9. Use of DNA mobility shift assay to characterize the sequence specificity of protein–DNA binding. If the protein binding to the labelled DNA sequence is capable of also binding to the unlabelled sequence it will do so and the radioactive retarded complex will be abolished.

Protocol 20. DNase I footprinting

1. Set up 100 μl binding reactions as for band-shift assay.[a,b] Incubate on ice for 40 min.
2. Dilute a 2 μg/μl stock of DNase I 1:100 immediately before use in buffer F (50 mM NaCl, 20 mM Hepes pH 7.9, 5 mM $MgCl_2$, 0.1 mM EDTA, 20% glycerol, 1 mM $CaCl_2$, 1 mM DTT). Add 1 μl to each sample and incubate at RT for a carefully timed 15–30 sec.

3. Stop the reaction by adding 100 μl 50 mM Tris–HCl pH 8.0, 2% SDS, 10 mM EDTA, 10 μg glycogen, 0.4 mg/ml proteinase K. Incubate at 37°C for 30 min, and then at 70°C for 2 min.

4. Extract the reaction with phenol:chloroform (1:1) and then with chloroform. And 15 μl 5 M LiCl and 600 μl ethanol and leave the samples overnight at −20°C to precipitate the DNA.

5. Spin down the DNA and resuspend it in 5–10 μl of sample buffer (950 μl formamide, 25 μl 1% bromophenol blue, 1% xylene cyanol). Load 20–50 counts per second in each well and run on a 6% denaturing polyacrylamide: bisacrylamide (19:1) gel containing 1 × TBE and 42% urea (w/v).[c]

6. Dry the gel under vacuum, and autoradiograph.

---

[a] In initial experiments it will be necessary to titrate the amount of extract used, the amount of poly dIdC and the magnesium concentration in order to obtain the appropriate level of digestion with DNAse I.

[b] In contrast to the DNA mobility shift assay where the probe can be labelled at both ends, the probe must be labelled at one end only. The probe is therefore prepared as in *Protocol 18* except that the labelled fragment is digested with a second restriction enzyme and gel purified to isolate a fragment with only one labelled end. Use approx. 5 fmol of probe per reaction labelled to 50–100 counts per second.

[c] The gel is pre-run for approx. 30 min prior to loading and then run at 1600 V/30 mA.

Hence the DNase I footprint offers an advance on the mobility shift assay, allowing a more precise visualization of the DNA protein interaction.

4.4 Methylation interference assays

The interaction between a DNA-binding protein and its specific DNA-binding site can be more precisely studied using the methylation interference assay in which the effect on the binding of the protein of methylating specific G residues in its binding site is assessed (28). This method therefore allows the precise assessment of the interaction of the DNA binding protein with individual nucleotides within its binding site. To do this, the DNA is partially methylated so that on average only one G residue per DNA molecule is methylated, and used in a standard DNA mobility shift assay. Following electrophoresis, the DNA which has bound protein and that which has failed to do so, are both excised from the gel, and their level of methylation at specific G residues compared by cleaving methylated Gs with piperidine (29). A lack of methylated G residues at a particular site in the protein-bound DNA indicates that methylation at this G blocks protein binding, and that it therefore plays a critical role in protein-binding (*Figure 12*).

Methylation interference can therefore be used as a supplement to DNase I footprinting by identifying the precise protein: DNA interactions within the footprinted region.

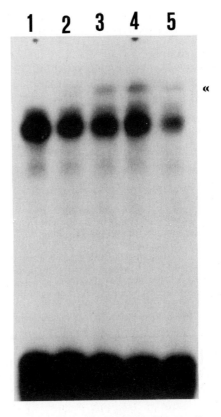

Figure 10. Identification of the nature of the additional neuronal-specific octamer binding protein detected in *Figure 8* by its reactivity with an antibody to the transcription factor Oct-2. *Tracks 2* to *5* show the result of band-shift assays with increasing amounts of the antibody whilst no antibody was added to the sample in *track 1*. Note the progressive appearance of a super-shifted band of low mobility (*arrowed*) as the antibody binds to the DNA-bound protein as well as the eventual inhibition of DNA binding by the protein. (For further details, see ref. 32.)

Protocol 21. Methylation interference

1. To prepare partially methylated probe, add end-labelled DNA to 200 μl 50 mM sodium cacodylate pH 8.0, 1 mM EDTA.
2. Chill on ice and add 1 μl dimethylsulphate (DMS). Incubate at 20°C for 3 min.
3. Add 2.5 μl 3 M sodium acetate pH 7.0 and 600 μl ethanol and incubate at −20°C overnight to precipitate the DNA.
4. Following centrifugation, discard the DMS containing supernatant into a 5 M NaOH solution. Wash the pellet with 70% ethanol, dry it and resuspend it in 10 μl of water.

5. Carry out a 120–200 μl DNA mobility shift reaction (see *Protocol 19*) using 50–100 fmol of partially methylated probe (approx. 4×10^5 c.p.m.).
6. Incubate the samples for 1 h on ice and load them on to 3–5 wells of a polyacrylamide gel prepared as for the DNA mobility shift assay.
7. After running, cover the gel with cling-film and expose it to X-ray film for approx. 4 h.
8. Excise the retarded and unretarded bands and extract the DNA with 1 M LiCl. Add 10 μg glycogen, phenol extract, and then ethanol precipitate the DNA.
9. Redissolve the DNA in 10 μl 1 M piperidine (freshly diluted) and heat it to 90 °C for 30 sec.
10. Cool the sample on ice, pulse-spin it in a microfuge, and freeze-dry the sample.
11. Redissolve the DNA in 50 μl of water and freeze-dry again.
12. Resuspend the DNA in 5 μl sample buffer (see *Protocol 20*), heat at 90 °C for 3 min and freeze it until ready to load on to a denaturing polyacrylamide gel.

4.5 South-western blotting

The techniques of DNA mobility shift, DNase I footprinting, and methylation interference discussed in previous sections can provide considerable information on the nature of the interaction between a particular DNA sequence and transcription factors. They do not, however, provide information on the protein itself and its characteristics. Ultimately, such information can be obtained by cloning the gene encoding the transcription factor using one of a number of different methods which are beyond the scope of this chapter (for discussion, see ref. 30). Prior to this, however, it is possible to determine the size of the protein by using the technique of South-western blotting (31). In this method, protein extracts are electrophoresed on a standard SDS–polyacrylamide gel and the separated proteins transferred to nitrocellulose membrane and probed with a radioactivity labelled oligonucleotide containing the DNA-binding site of interest. A protein capable of binding to this specific site will do so producing a radioactive band, and its size can be determined by comparison to marker proteins of known size.

Protocol 22. South-western blotting

1. Electrophorese approx. 50 μg of protein extract on a standard SDS–polyacrylamide gel and transfer on to a nitrocellulose membrane at 20 mA overnight in 25 mM Tris–HCl, 200 mM glycine, 20% methanol.

Regulation of gene expression in neuronal cell lines

Protocol 22. *Continued*

2. Denature the protein bound to the filter in 6 M guanidine-hydrochloride for 30 min.
3. Renature the protein overnight in 10 mM Hepes pH 7.9, 100 mM KCl, 0.1% NP40, 1 mM DTT.
4. Incubate the filter in blocking buffer (10 mM Hepes pH 7.9, 1 mM DTT, 5% non-fat dried milk) for 60 min at RT with gentle agitation.
5. Incubate the blot for at least 2 h (or overnight) in 10 mM Hepes pH 7.9, 50 mM NaCl, 0.1 mM EDTA, 1 mM DTT, 0.25% non-fat dried milk, and about 3.5×10^7 c.p.m. ml of ^{32}P-labelled concatamerized oligonucleotide probe.[a]
6. Wash the blot in 10 mM Tris–HCl pH 7.5, 50 mM NaCl; wrap it in cling-film and autoradiograph.

[a] Probe is labelled as described in *Protocol 18*.

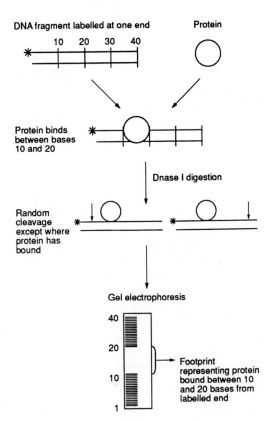

Figure 11. DNase I footprinting assay in which the region bound by a protein is identified by its resistance to digestion by DNase I.

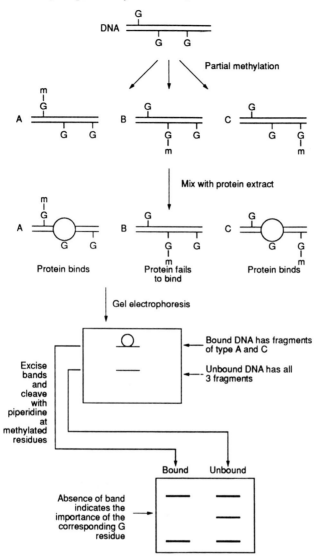

Figure 12. Methylation interference assay in which the guanine residues which are critical for binding of a protein to the DNA can be identified by the fact that their methylation prevents protein binding and hence the formation of a retarded complex in a DNA mobility shift assay.

5. Conclusion

The methods described in this chapter will allow the characterization of the processes which regulate the expression of a specific gene, the DNA sequences which are important in these processes, and the transcription factors which bind to them. The successful execution of these techniques, however, is critically dependent on the availability of neuronal cell lines which provide a source of large amounts of homogeneous material for extract preparation, transfection, etc. Ultimately, however, insights obtained in this way can be applied to the study of gene regulation in neuronal cells *in vivo* where the amounts of material are much more limited. Thus, for example, if a particular transcription factor is identified in neuronal cell lines as playing an important role in the regulation of a particular gene, the expression of the mRNA encoding the factor can be studied in neuronal cells *in vivo*, using either *in situ* hybridization or the PCR, in order to ascertain whether it is expressed in the cells expressing the gene whose activity it may regulate. In this way, the study of gene expression in neuronal cell lines can greatly aid an understanding of the processes which regulate gene expression in neuronal cells *in vivo*.

References

1. Latchman, D. S. (1990). *Gene regulation*. Unwin-Hyman, London.
2. Chapman, A. B., Costello, M. A., Lee, F., and Ringold, G. M. (1984). *Mol. Cell. Biol.*, **3**, 1421.
3. Lehrach, H., Diamond, D., Wozney, J. M., and Boedtker, J. (1977). *Biochemistry*, **16**, 4763.
4. Feinberg, A. P. and Vogelstein, B. (1983). *Anal. Biochem.*, **132**, 6.
5. Zinn, K., Dimaio, D., and Maniatis, T. (1983). *Cell*, **34**, 365.
6. Latchman, D. S. (1990). *New Biologist*, **2**, 297.
7. Kawasaki, G. S. (1990). In *PCR protocols. A guide to methods and applications* (ed. M. A. Innis, D. M. Gelford, J. J. Sninsky, and J. J. White), p. 21. Academic Press, San Diego.
8. Higuchi, R., Krummel, B., and Saiki, R. K. (1988). *Nucleic Acids Res.*, **16**, 7351.
9. Mason, I. J., Murphy, D., Munke, M., Francke, U., Elliot, R. W., and Hogan, B. L. M. (1986). *EMBO J.* **5**, 1831.
10. Patel, R., Chan, W. L., Kemp, L. M., La Thangue, N. B., and Latchman, D. S. (1986). *Nucleic Acids Res.*, **14**, 5629.
11. Gorman, C. M. (1985). In *DNA cloning: a practical approach* (ed. D. M. Glover), Vol. 2, p. 43. IRL Press at Oxford University Press, Oxford.
12. Herbomme, I. P., Bourachot, D., and Yaniv, M. (1984). *Cell*, **39**, 653.
13. de Wet, J. R., Wood, K. V., De Luca, M., Helinski, D. R., and Subramani, S. (1987). *Mol. Cell. Biol.*, **7**, 725.
14. Frebourg, T. and Brison, O. (1988). *Gene*, **65**, 315, 15.
15. Serfling, E., Jason, M., and Schaffner, W. (1985). *Trends Genetics*, **1**, 224.
16. Queen, C. and Baltimore, D. (1983). *Cell*, **33**, 741.
17. Potter, H., Weir, L., and Leder, P. (1984). *Proc. Natl. Acad. Sci. USA*, **81**, 7161.
18. Bradford, M. (1976). *Anal. Biochem.*, **72**, 248.
19. Sheng, M., Dougan, S. T., McFadden, G., and Greenberg, M. E. (1988). *Mol. Cell. Biol.*, **8**, 2787.
20. Luckow, B. and Schatz, G. (1987). *Nucleic Acids Res.*, **15**, 5490.
21. Latchman, D. S. (1990). *Biochem. J.*, **270**, 281.

22. Manley, J. L., Fine, A., Cano, A., Sharp, P. A., and Gefter, M. L. (1980). *Proc. Natl. Acad. Sci. USA,* **77,** 855.
23. Dignam, J. D., Lebovitz, R. M., and Roeder, R. G. (1983). *Nucleic Acids Res.,* **11,** 1575.
24. Fried, M. and Crothers, D. M. (1981). *Nucleic Acids Res.,* **9,** 6505.
25. Garner, M. M. and Revzin, A. (1981). *Nucleic Acids Res.,* **9,** 3047.
26. Galas, D. and Schmitz, A. (1978). *Nucleic Acids Res.,* **5,** 3157.
27. Dynan, W. S. and Tjian, R. (1983). *Cell,* **35,** 79.
28. Siebenlist, U. and Gilbert, W. (1980). *Proc. Natl. Acad. Sci. USA,* **77,** 122.
29. Maxam, A. M. and Gilbert, W. (1980). *Methods in enzymology* (ed. L. Grossman and K. Moldave), Vol. 65 p. 499. Academic Press, Orlando, Florida.
30. Latchman, D. S. (1991). *Eukaryotic transcription factors.* Academic Press, London.
31. Sambrook, J., Fritsch, E. F., and Maniatis, T. (1989). *Molecular cloning—A laboratory manual.* Cold Spring Harbor Laboratory Press, Cold Spring Harbor, NY.
32. Dent, C. L., Lillycrop, K. A., Estridge, J. K., Thomas, N. S. B., and Latchman, D. S. (1991). *Mol. Cell. Biol.,* **11,** 3925.
33. Wheatley, D. S., Dent, C. L., Wood, J. N., and Latchman, D. S. (1991). *Exp. Cell Res.,* **194,** 78.

9

Cell lines in developmental neurobiology: assays of adhesion and neurite outgrowth

JOHN A. PIZZEY

1. Introduction

Communication between cells is a fundamental requirement for the diverse interactions implicit in normal development and tissue homeostasis. Such communication is generally mediated by membrane-associated molecules which can be classified (1) into three broad functional classes; cell adhesion molecules (CAMs), substrate adhesion molecules (SAMs), and cell junctional molecules (CJMs). However, it should be noted that these are not mutually exclusive classifications; for example, the neural cadherin (N-cadherin) is present both on the surface of neural cells as a CAM (2), but is also found in the adherens junctions on some non-neural tissues (3) and thus could be considered a junctional molecule. Although some overlap exists between these functional groupings, most attention will be given to CAMs wherever adhesion molecules are mentioned in this chapter. These molecules appear to be involved in cellular processes such as adhesion and recognition which, until fifteen years ago, were frequently described in phenomenological terms. However, largely by the application of immunological and molecular biological techniques, the identity of many of these molecules is now known. Furthermore, a range of functional assays is now available for the identification of these adhesive mechanisms and the molecules involved. After consideration in Section 2 of some basic principles and requirements for measuring intercellular adhesion, assays that are used for these types of analyses are described, with special reference to neural cell lines, in Section 3.

Although recognition and adhesion between cells of the same type is likely to be a general requirement for normal tissue development, the neural system presents unique problems for cell–cell interactions. For example, in addition to the formation of neuronal synapses in the central nervous system (CNS) and neuromuscular junctions in the periphery, neurons have to extend their axons over very great distances and through a diverse tissue milieu in order to find their correct targets. Several adhesion molecules have been shown to be directly involved in neurite outgrowth and have additionally been implicated in synaptogenesis. Assays for detecting and quantitating such aspects of neuronal differentiation are considered in Section 4. Finally, in Section 5, I will briefly consider future prospects for further elucidation of the function and nature of the molecules involved in promoting adhesion or neurite outgrowth.

2. Intercellular adhesion

2.1 Basic principles

2.1.1 Dissociation methods

An obvious and fundamental prerequisite for testing the ability of cells to adhere *in vitro*, by either specific or general mechanisms, is to ensure that the molecules involved are present and active at the time of the assay. Almost all cell adhesion molecules known to date are surface-associated. An exception to this may be one of the various isoforms of the neural CAM (N-CAM). It has been reported that a secreted (soluble) variant of this CAM may exist (4) although evidence exists to suggest that it may not act as an intercellular CAM *in vitro* (5). However, in general, the need to dissociate a monolayer culture prior to any assay will necessarily perturb the cell surface. Some of the more common methods of dissociation for adhesion assays using cultured cells are given in *Table 1*.

Table 1. Methods of cell dissociation for adhesion assays

Method of dissociation	Advantages	Disadvantages
Mechanical	No chemical perturbation	Low yield of viable cells
		Cell monolayers may be sheared
		Irregular cell shape
Trypsinization	Rapid and efficient	Removal or inactivation of some classes of adhesion molecules
	Ease of standarization	
Calcium chelation	No enzyme perturbation	Not suitable for all cells
	Ease of standardization	Inactivation of some classes of adhesion molecules
Ice cold buffer	No chemical perturbation	Not suitable for strongly adherent cells
		Temperature shock
		Potential change of molecular conformations

All of the procedures have advantages and disadvantages and although removal of cells from a substratum using a rubber policeman introduces no chemical change to the cell surface, it has long been known that this method can shear a monolayer in different planes, resulting in both mechanical disruption of the cell membranes and a low yield of a viable, homogenous population (6). Furthermore, such cells are often detached from the substratum in sheets or clumps, and additional mechanical disruption, such as repeatedly passing the suspension through narrow-gauge syringe needles (see below) is frequently required to yield single cells. Finally, cells detached by such procedures are very heterogeneous in their shape, and this can also affect the reproducibility of subsequent adhesion assays or even prohibit their use. The need for a population of cells of even shape is met by all the other means of dissociation described; if cells can be removed by trypsinization, calcium chelation, or ice-cold temperatures, then they will round up into spheres. Although this is clearly an unphysiological state for the cells, it allows for standardization of the assay conditions and methods can be employed to establish recovery of cell membrane components which may be lost or inactivated by these methods (see Section 2.1.2). Dissociation of monolayers by calcium chelation or incubation in buffers at 4°C does not consistently yield large numbers of cells in suspension for all lines, and the latter technique is particularly inefficient for very adherent cells.

In consideration of these factors, it has generally become accepted that limited proteolytic digestion (typically using trypsin, although collagenase or hyaluronidase may also be used) is the desired mode of dissociation even when the cellular function under study involves surface-mediated events. However, it is important to reiterate here that, irrespective of the dissociation method used, it is usually necessary to ensure that the cell population under study is of uniform size and shape, a consideration that is fulfilled when using cell lines. Additionally, it is essential to confirm that the final suspension contains only single cells and not doublets, triplets and multicellular aggregates. Often this is achieved by repeated pipetting or by passing the suspension through a narrow-gauge syringe. However, this latter method should only be used on cells which are relatively robust, and should be augmented by passing the suspension through a sterile filter of suitable pore size under gravity. Confirmation of the monocellular identity of the population should be performed by visual inspection using haemocytometers or by the use of electronic particle counters, such as Coulter counters (Coulter Instruments, Luton).

2.1.2 Ca^{2+}-dependent and Ca^{2+}-independent adhesion

From studies of neural (7), myogenic (8, 9), and other (10) cell types it has become clear that the method of cell dissociation can determine the class of adhesive molecules which remain active on the cell surface. Briefly, mechanisms of intercellular adhesion can often be functionally divided into Ca^{2+}-

dependent (CD) and Ca^{2+}-independent (CI) categories (1). Jessell, 1988 (11), provides a good review of CAMs within the neural system, but briefly the major CD adhesion molecule in the nervous system known to date is N-cadherin (for reviews, see refs 2 and 12) while several of the neural CI adhesion molecules are members of the immunoglobulin superfamily (13). Of this latter group, perhaps N-CAM, L1 (also known as NILE or Ng-CAM) and Po are among the best characterized examples (see, for example, refs 14–16). Dissociation procedures can be chosen such that either CD or CI mechanisms are operative and these are described in *Protocols 1* and *2*. It should be noted that some of the steps in these procedures need to be determined empirically for individual cell lines. For example, some lines may be particularly sensitive to trypsinization; in such cases either the incubation time or the concentration may be reduced. However, it should be noted that the concentration of trypsin used (in the presence of EGTA) is fundamental in determining the adhesive mechanisms which remain active. The concentration of trypsin stated in *Protocol 1* will leave CI adhesive systems functional in embryonic chick neural retina cells (7) and myoblasts (8, 9). However, the form of trypsin used markedly affects its purity and therefore its catalytic activity. The concentration given in *Protocol 1* refers to crystalline trypsin, which is much purer than the lyophilized form. For example, if lyophilized trypsin is used, a concentration of 0.05% in HBSS is sufficiently low to inactivate the CD component (only) of mouse C2 myoblast adhesion (17). If the trypsin concentration is too high, both CD and CI adhesive mechanisms will be inactivated. It has been shown for chick retinal cells that 68 National Formulary Units (NFU) of trypsin in 1.25 ml 1.3 mM EGTA will not affect the CI component of the adhesiveness of these cells, whereas 3000 NFU of trypsin will inactivate both CD and CI systems (7). If 3X crystalline porcine trypsin is used, these values are equivalent to 0.001 and 0.044% respectively. Alternatively, if it is desired that both CD and CI mechanisms remain operative, trypsin should be omitted from step 3 of *Protocol 1*. However, although this procedure does not appear functionally to damage CD adhesion (8), it is often difficult to achieve complete dissociation in EGTA alone. Therefore, an alternative strategy that leaves both systems intact is to dissociate the cell monolayers in a low concentration of trypsin in the presence of Ca^{2+} ions. This can be achieved by replacing EGTA in *Protocol 1* with 2.5 mM $CaCl_2$. In this case, $CaCl_2$ should be added to HBSS 10–15 min before the addition of trypsin.

Protocol 1. Dissociation protocol to identify CI adhesive mechanisms

1. Grow the cultures to 70–100% confluency in 150-mm culture dishes.
2. Wash the cultures in 10 ml Ca^{2+} and Mg^{2+}-free Hanks' balanced salt solution (HBSS) containing 1.3 mM EGTA, pH 7.4 (HE).
3. Remove the HE and replace this with 5.0 ml HE containing 0.001% trypsin,[a] pre-warmed to 37°C.

4. Incubate the cells at 37°C with gentle agitation for 20 min.
5. Inactivate the trypsin by the addition of leupeptin to a final concentration 5 μg/ml, and collect the cell suspension into 20 ml HE at 4°C. Ensure that >95% of cells have been removed.
6. Pellet the cells by centrifugation at 400 g for 10 min at 4°C.
7. Resuspend cells in 20 ml HE at 4°C and centrifuge as before.
8. Resuspend cells in 2 ml HE at 4°C and separate single cells from doublets, triplets, and small aggregates by passing repeatedly through a 21-gauge syringe needle and then through a sterile 20 μm Nitex filter.
9. Keep the suspension on ice while determining the cell concentration by haemacytometry or by using an electronic particle counter.

[a] This concentration refers to 3X crystalline trypsin.

Protocol 2. Dissociation protocol to identify CD adhesive mechanisms

1. Grow the cultures to 70–100% confluency in 150-mm culture dishes.
2. Wash the cultures in 10 ml Ca^{2+}- and Mg^{2+}-free HBSS containing 2.5 mM $CaCl_2$, pH 7.4 (HC).[a]
3. Add trypsin (see footnote a to Protocol 1) to 3 ml HC such that the final concentration is 0.044%[b] and pre-warm to 37°C.
4. Aspirate HC from cell cultures and incubate in trypsin solution for 20 min at 37°C with gentle agitation.
5. Continue from step 5 in Protocol 1, but substitute HC for HE throughout.

[a] This is prepared by adding 0.5 ml 50 mM $CaCl_2$ to 9.5 ml HBSS.
[b] It has been reported (9) that Ca^{2+} must be added to the HBSS at least 10–15 min before the trypsin.

2.1.3 Recovery of adhesiveness

Whichever mode of dissociation is used, some damage to the cell surface molecules of interest is almost inevitable. However, under appropriate conditions living cells can synthesize and translocate the lost or damaged molecules back to the cell surface. This may not be necessary, or even desirable, if a specific class of CAMs have been functionally selected for using the dissociation protocols described above. On the other hand, if the total adhesive properties of the cells are of interest, then a recovery period should be introduced during which time the full surface properties, or an acceptable level of them, are restored. Different strategies have been used to allow for such restoration and involve recovery either on non-adhesive substrata or in suspension, and these are described below.

The most common method for recovery of cell adhesiveness is to wash the cells after dissociation and then resuspend them in a supporting medium which is gently agitated to prevent the cells from settling to the bottom of the tubes. The most important factors to consider are:

- the selection of the recovery medium;
- means to obviate mechanical damage; and
- avoidance of re-aggregation.

The choice of medium may be restricted by the requirements of the individual cell type. If the cell population can survive for even a limited time in defined medium, then this obviates several problems associated with other media. First, cells do not survive for very long in unsupplemented minimal media without serum. Second, although the supplementation of serum promotes survival and recovery, it contains many adhesive proteins which can become adsorbed to the cell surface and either mask or augment the adhesiveness due to endogenous CAMs. To some extent, this can be overcome by using fibronectin-depleted serum, although it must be remembered that this only removes one component, albeit a major adhesion molecule, from the recovery medium. I have found that Sato medium, which is suitable for use in culturing various neural cell lines (ref. 18 and Chapter 3), is an ideal first choice for the recovery phase of any adhesion assay. A suitable recovery procedure using Hepes-buffered Sato and containing leupeptin, which I have used for C2 myoblasts, mouse L cells, and 3T3 fibroblasts is given in *Protocol 3*.

Protocol 3. Recovery of cells in Sato medium after dissociation

1. Dissociate the cell monolayers according to desired protocol, add leupeptin (final concentration 5 μg/ml) if trypsin is used, and centrifuge at 400 g for 10 min at 4 °C.
2. Resuspend the cell pellet in 10 ml HBSS and centrifuge again at 4 °C for 10 min.
3. Resuspend the cell pellet in 2–3 ml at a concentration of approximately 2×10^5 cells in modified Sato medium[a] and transfer to a 15-ml conical screw-cap tube.
4. Fill the tube to capacity with modified Sato medium and screw on the cap (filling the tube to capacity is important, since this will leave only a small air bubble in the tube which will minimize liquid turbulence, and therefore mechanical damage, to the cells during the recovery phase).

5. Place the tubes on an orbital rotator (< 30 r.p.m.) and incubate at 37 °C.
6. After a recovery period (typically 2–24 h), centrifuge the cells at 400 g for 10 min at 4 °C.
7. Wash the pellet in 10 ml HBSS at 4 °C and centrifuge as before.
8. Resuspend the cells in aggregation medium at the desired concentration for an adhesion assay.

[a] 1 litre of modified Sato medium represents Dulbecco's MEM/10 mM Hepes containing: 0.062 mg progesterone; 16.1 mg putrescine; 0.4 mg thyroxine; 0.039 mg selenium; 0.338 mg triiodothyronine; 10.0 mg bovine pancreas insulin; 100 mg transferrin; 5.0 mg leupeptin; 1% (v/v) Path-O-cyte in bovine serum albumin solution (from Miles Chemical Company).

An alternative strategy for recovery of cell surface properties involves the use of non-adhesive substrates. In this case, a common choice of substratum is the surface of sterile bacteriologic grade (i.e. not tissue culture grade) dishes. The dissociated cell suspension is washed and seeded at approximately 2×10^5 cells/ml in 150-mm dishes. After the appropriate recovery time, the medium is harvested and the plates gently washed with fresh medium to remove the loosely-attached cells. Both fractions are then pooled and centrifuged. Again, a choice of recovery media is available, although Sato is a good starting point for this recovery procedure also, although here, too, fibronectin-depleted serum has previously been used successfully.

There are two final points to note with regard to recovery phases in adhesion assays. First, the forces which are present even in gently swirling medium may still be too great for extremely fragile cells. This is more of a problem when dealing with primary cell cultures but may also extend to some cell lines. In such cases, the investigator has no choice but to allow the cells to recover on a substratum. Secondly, the density of cells in the recovery medium is critical and although the value given in *Table 1* is appropriate for mouse C2 and L cells, it may have to be modified for other cell lines. The importance of cell density is due to the fact that if the cell concentration is too high (either in suspension or substratum-based recovery methods), then the probability of cell–cell adhesions is increased. These adhesions become very stable and either introduce pre-formed aggregates into subsequent adhesion assays or lead to a substantial decline in cell titre if such aggregates are removed by filtration (as they should be) prior to assay. On the other hand, if cell density is too low, then there may be insufficient endogenous factors which are required for the maintenance of most cell populations in defined media. Additionally, at low cell densities, the numbers of individual flasks or dishes required to provide the total number of cells for assay may be too great to handle easily, especially if multiple treatment or different cell lines are to be assayed simultaneously.

3. Intercellular adhesion assays
3.1 Basic principles
A great variety of assays exist for measuring adhesive cellular interactions, but they can generally be divided into those which measure adhesive and recognition events between cells and those which analyse the interactions between cells and a substratum. The latter type of measurements invariably involve seeding cells on to a surface coated with relevant protein (either purified or in crude extract form) and determining the cell population's ability to attach, spread, and, where appropriate, differentiate and/or locomote. Such assays can be quantitated by visually counting the number of attached cells or by measurements of cell area, or cell motility. Alternatively, attached cells can be quantitated by immunological techniques such as enzyme-linked immunosorbent assays (ELISA), which are discussed in Section 4, or by prior radioisotopic labelling (19). However, in this section I shall concentrate on those assays which measure recognition and adhesion between cells rather than between cells and a non-cellular substratum.

Most intercellular adhesion assays can also conveniently be divided into two further categories; those which measure the formation of multicellular aggregates and those which monitor the disruption of pre-formed adhesions. However, the latter method often provides data which are more difficult to interpret and also probably involve more damage to the cell surface (20). A sample of some of the great number of available intercellular adhesion assays, together with some of their advantages and drawbacks or limitations, are listed in *Table 2* and discussed in detail below.

3.1.1 Aggregation on rotators
By far the commonest form of adhesion assay involves aggregation of a single cell suspension by subjecting it to agitation on various forms of rotators set at different speeds. Increasing speed of rotation will increase the number of collisions between cells and therefore the opportunity for adhesion molecules to react with their targets. However, greater rotational speeds will also increase the probability of the same adhesions being disrupted by the fluid shear exerted upon them. In general, speeds of between 30 and 90 r.p.m. are used in this form of assay, although equivalent assays have also been performed when the suspension is subjected to no fluid shear, i.e. the assay is conducted in still medium (21, 22). Since cell adhesion is a multi-step phenomenon in which initial recognition events are followed by mechanisms by which these early adhesions are strengthened (23, 24), this latter application may be particularly suitable for assaying adhesion due to molecules which may be involved in the recognition phase of cell adhesion (5).

The method described in *Protocol 4* is useful for measuring adhesion in a range of cell types including neural, myogenic and fibroblastic cells. How-

Table 2. Methods of measuring intercellular adhesion

Method of assay	Advantages	Disadvantages
Aggregation on rotators	No specialist equipment required	No quantitative estimate of cell adhesiveness
	Quantitative estimate of aggregation kinetics possible	
Aggregate in Couette viscometers	Quantitation of cell adhesiveness possible by calculations of collision efficiencies	Specialist equipment needed
	Quantitative estimate of aggregation kinetics possible	Cells may be distributed unevenly throughout medium during long assays
Aggregation in multi-channel aggregometers	Quantitative estimate of aggregation kinetics possible	Specialist equipment needed
		No quantitative estimate of cell adhesiveness
Disruption of adhesion by centrifugation	Quantitative estimate of cell adhesion possible	Laborious and relatively expensive
	Storage of experiments possible prior to analysis	Possible false readings due to autolysis
	Multiple samples can be processed simultaneously	

ever, if the contribution of either a particular class of adhesive mechanism (possibly as isolated by the dissociation procedures described in Section 2.1.1) or an individual CAM (as a result, for example, of cDNA transfection strategies) is to be investigated, then varying the r.p.m. to be used may be essential (5). As indicated in the notes to *Protocol 4*, the choice of aggregation medium may depend on the requirements of the cell type. In principle, a good choice of medium is usually the medium which the cultured cells are grown in, without the inclusion of serum. This will generally limit the time for which the assay can be performed to approximately 1–2 h but this should be more than sufficient for equilibrium to be reached. In this case, equilibrium is defined as the aggregates reaching a final maximum size as the rate at which cells are recruited into the aggregates is equal to the rate at which they are shed from them. Additionally, some groups advocate the inclusion of 0.01% DNase I in the aggregation medium. However, this should not be necessary if lysis during preparation and the adhesion assay is kept to a minimum. I have found that the addition of DNase I makes no significant difference to the aggregation kinetics of several cell types.

Protocol 4. Measurements of cell adhesion using laboratory rotators

1. Dissociate the cells according to the desired protocol and allow to recover if necessary.
2. Centrifuge the cells at 400 g for 10 min at 4°C.
3. Wash the cells in 10 ml HBSS at 4°C and centrifuge as before.
4. Resuspend the cells in 2 ml modified Eagle's medium (MEM)[a] and gently pass the suspension twice through a 21-gauge syringe needle.
5. Add 3 ml MEM and filter through a 20 μm Nitex membrane.
6. Count the cells either by haemocytometry or by taking a 50 μl aliquot and mixing with 10 ml PBS and counting in an electronic particle counter.[b]
7. Bring the cells to a concentration of 1×10^6 cells/ml in MEM.
8. Transfer 1.5-ml aliquots into gelatin- or silicon-coated Eppendorf tubes and place on a rotator at 90 r.p.m. at 37°C.[c]
9. Remove 50 μl aliquots at 15-min intervals and count the cells as at step 6.

[a] The presence of Ca^{2+} may depend on the requirements of the particular cell type and the adhesive mechanisms under study (see text). However, for short-term assays, other (non-adhesive) components of the aggregation medium may be less critical.

[b] Both methods of counting have advantages; haemocytometry requires no special equipment and allows a visual inspection of the quality of the cell suspension. Electronic particle counting allows for multiple samples to be counted rapidly but assumes that all cells are viable. Alternatively, cells can be counted by flow cytometry (62).

[c] According to the temperature at which the assay is conducted, different aspects of cell adhesion can be identified. For example, it appears that, in several different cell lines, the initial intercellular recognition events can occur at low temperatures while it is at 37°C that strengthening of adhesion occurs (23, 63, 64). See text for further details.

At different time points of incubation, aliquots of the cell suspension are taken and counted. Again, as described in the notes to *Protocol 4*, a choice of counting methods are available. Whichever method is chosen, the viability of the cell suspension should be monitored at regular intervals during the course of the assay. The simplest method for assessing viability is to mix an aliquot of the cell suspension with an equal volume of 1% trypan blue and the proportion of viable cells are then estimated from the fraction which excludes the dye. It should be remembered that this only provides an approximation of the proportion of viable cells and that many cells may exclude the dye but still have damaged membranes. However, any indication that > 5% of the cell population take up the dye should be interpreted as representing unacceptable cell death and appropriate measures (less robust dissociation methods, slower speeds of rotation for cell aggregation, inclusion of supplements in the aggregation medium, etc.) should be taken.

Several different indices which reflect cell adhesion can be measured. First, total particle number can be counted. This is probably the commonest form of measurement and is based on the fact that as cells adhere to each other, multicellular aggregates are formed and thus an increasingly greater number of single cells are lost from the suspension. The adhesiveness of the cell population can be related to the rate of aggregation, and the extent of aggregation can be represented by the index $(N_0-N_t)/N_0$, where N_t is the total number of particles (both single cells and aggregates) after incubation time t and N_0 is the total particle number at the start of incubation (*Figure 1*).

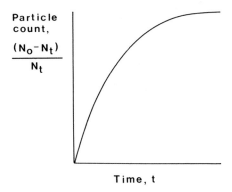

Figure 1. Kinetics of cell aggregation reflecting adhesive interactions between cells in suspension. This curve is characteristic for rapidly adhering cells which soon reach equilibrium and follow first-order kinetics (see text). If the total number of particles counted after time t is N_t, and the total number of particles at the start of the assay is N_0, then the proportion which have been recruited into aggregates is given by: $(N_0 - N_t)/N_0$. If first-order kinetics are followed, then plotting log (N_t/N_0) against t will yield a straight line.

Alternatively, the proportion of either aggregates or single cells can be measured separately for each time-point. If an electronic particle counter is used, this can be refined further by setting the size windows such that either only aggregates of a particular size class are measured or the whole spectrum of particle size classes can be determined over the course of the assay.

3.1.2 Aggregation in Couette flow

Measurements of cell adhesion by monitoring aggregation in a swirling medium (as described in Section 3.1.1) provide a semi-quantitative estimate of the aggregation kinetics of the cells under study but do not allow for a fully quantitative index of their total adhesiveness. In most instances this is not a serious drawback, but in other cases it provides a distinct advantage. For example, it does not allow for adhesive interactions between two morphologically similar cell populations to be measured without immunological identification or prior labelling of one of them (25). The failure to allocate a quantitative index of 'adhesiveness' is due to the fact that the pattern of liquid flow in such shaken suspensions is extremely complex and the shear rate, upon which aggregate formation depends, cannot be determined. To provide a constant, known laminar shear flow of liquid, Curtis introduced the use of the Couette viscometer to provide for more accurate measurements of cell adhesion (25). Couette viscometers are available from suppliers such as Fisons (Loughborough) or Contravenes (Ruislip). The theoretical basis of the use of Couette viscometry to measure cell adhesion is described in detail elsewhere (25–27) but, essentially, it consists of an outer hollow cylinder which rotates at fixed, pre-selected speed around a static, solid cylinder. The cell suspension is introduced into the gap between the two cylinders, and the outermost of the two then exerts a laminar shear flow upon the dissociated cells under carefully controlled microenvironmental conditions. This allows for calculations to be made of both the total number of collisions in a cell suspension and the number of these collisions which result in adhesions over a defined time-period.

As mentioned in the preceding section, many individual CAMs will have different adhesive strengths and the conditions of any adhesion assay often have to be varied to identify them. In Couette viscometry, varying the laminar shear force is equivalent to changing the r.p.m. in the rotational assays described above. However, in this case the shear force can be calculated, and this in turn permits the collision efficiency (α) to be calculated, which can be regarded as a quantitative index of the 'adhesiveness' of the cell type.

The shear rate is calculated from:

$$G = \omega \left\{ \frac{r_o^2 + r_i^2}{r_o^2 - r_i^2} \right\}$$

where G is the shear rate in $\sec^{-1}$, ω is the angular velocity of the rotating cylinder, and r_o and r_i are the radii of the outer and inner cylinders respectively. I have found that a shear rate of $10 \sec^{-1}$ is suitable for a wide range of cell lines and primary cell cultures (17, 27) although this value may have to be altered substantially for cells which are particularly unadhesive or aggregate very rapidly. More details on the calibration of the assay are given in ref. 26.

Cells are prepared for assay in the same way as that described in *Protocol 4* and 1.0 ml is transferred to the well of the Couette viscometer. However, for this assay the aggregation medium should contain 5% Ficoll. This supplement is necessary to prevent cells from becoming unevenly distributed throughout the medium during prolonged assays. Furthermore, since this method represents an open system, it is essential that the medium is suitably buffered. Typically, 20-μl aliquots are taken at 5-min intervals and the total number of particles are counted. Again, as cells adhere and aggregate, this is reflected as a decrease in this value. For each time-point, the collision efficiency coefficient α can then be calculated as follows:

$$\alpha = \frac{\pi \ln (N_t/N_0)}{4G\phi t}$$

where N_t and N_0 are the total particle counts at individual sample times and at the start of the incubation respectively, ϕ is the volume fraction of the cells (which should be < 0.1%), t is time in seconds, and G is the shear rate as defined above. If aggregation follows first-order kinetics then while the assay is in this range, each value for α (determined at different time-points in any one assay) should be approximately the same. Thus, the mean value can be determined for the collision efficiency of the cells in that particular assay. This provides a quantitative measure of adhesion which allows for a fuller and simpler statistical treatment of data.

3.1.3 Disruption of adhesion by centrifugation

A different strategy to measure cell adhesion was developed by McClay et al. (23) who designed an assay whereby quantitative measurements of cell binding strength could be made. The assay is shown diagrammatically in *Figure 2*, and described in *Protocol 5*. This assay has a number of advantages over

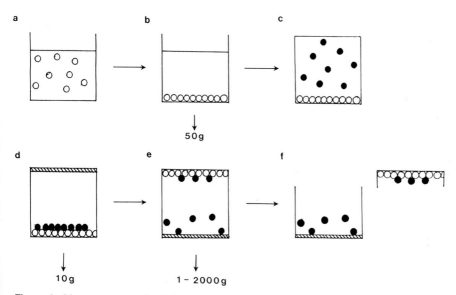

Figure 2. Measurements of cell adhesion by centrifugation as described by McClay et al. (23). Target cells are introduced into the wells of microtitre plates (a) and centrifuged to bring them into contact with the PLL-coated surface (b). Labelled probe cells are then introduced into the wells which are then filled to capacity (c). The wells are sealed and centrifuged to bring the probe cells into contact with the unlabelled cell monolayer (d). The plates are then incubated, inverted, and subjected to an appropriate centrifugal force which will tend to remove the probe cells (e). Plates are then frozen (still inverted) on dry-ice, the bottom 3 mm of each well are cut off and placed directly into scintillation vials for counting (f).

others mentioned here, including the ability to process a very large number of samples simultaneously. This assay has also been used to provide quantitative support for the hypothesis that the initial recognition events in cell adhesion are mediated by weak interactions and are temperature-independent (23), in contrast with the more traditional assays which are performed at 37°C. However, it also presents certain disadvantages (*Table 1*).

Briefly, cells are labelled isotopically and brought into contact by gentle centrifugation with cell monolayers established in the wells of microtitre plates. As for all the assays described here, it is important to recognize the potential need for a recovery period following dissociation of both target (monolayer) and probe cells. In the present assay, the target cells are initially gently spun on to a poly-L-lysine (PLL) substratum to prevent them being dislodged at later stages of the assay. Thus these cells can be allowed to recover before spinning down on to PLL, or alternatively step 4 in *Protocol 5* can be extended to introduce a longer recovery phase at this point.

Protocol 5. Quantitative measurements of cell adhesion by centrifugation[a]

Preparation of monolayer
1. Coat the wells of tissue culture treated flat-bottom microtitre plates with 50 μg/mg poly-L-lysine (M_r 60 000) for 1 h at 24°C.
2. Wash the wells twice with sterile PBS and once with culture medium.
3. Dissociate the target cells according to the desired protocol and seed the cells into wells in 100 μl of culture medium and spin the plate at 50 g for 3 min.[b]
4. Incubate the plates for 1 h at 37°C.[c]
5. Block non-specific binding of probe cells by incubating the plates with 10 μg/ml poly-L-glutamic acid (M_r 30 000) in PBS for 30 min at 37°C.
6. Remove the poly-L-glutamic acid, wash twice with culture medium and proceed to step 11.

Probe cell preparation
7. Wash the probe cell monolayers twice with PBS and once with leucine-free medium.
8. Label the monolayers with [3H]-leucine (10 μCi/ml) in leucine-free medium for 16 h at 37°C.
9. Wash the cells twice with PBS and dissociate and wash them using the desired protocol. If a recovery period is introduced, the presence of [3H]-leucine should be maintained, followed by washing twice in PBS.
10. Resuspend the cells in culture medium at 4°C at a concentration of 1.0×10^6 cells/ml.

Determination of probe–target interactions

11. Add 100 μl of probe cell suspension to each well and then add a further 200 μl of culture medium.[d]

12. Seal the wells by rolling on an adhesive microtitre plate sealer (Dynatech).

13. Bring the probe cells into contact with target cells by centrifuging at 10 g for 10 min at 4°C.[e]

14. Incubate the plates for the desired time and temperature. Conditions chosen here will help to determine whether recognition or secondary strengthening adhesions are being measured.

15. Invert the plates and centrifuge once more for 10 min at 4°C.[f]

16. Place the microtitre plates (still sealed and inverted) on to dry ice.

17. When frozen, clip off the bottom of each well, containing the monolayers, attached probe cells and an aliquot of the frozen sample medium, and transfer them to scintillation vials. It is important to cut each well at the same point; cutting approx. 3 mm from the base of each well will remove approx. 40 μl of frozen medium.

18. Add an appropriate scintillation cocktail and count the samples.

[a] After ref. 23.
[b] The number of cells needed to establish a confluent or near-confluent monolayer has to be determined empirically since it will naturally depend on the size of the cells to be used.
[c] A recovery phase can be introduced here by incubating the monolayer for a longer period (see text).
[d] The medium level will rise slightly above the level of each well. This ensures that air bubbles will not form when the tapes are sealed. Contamination by small amounts of suspensions from neighbouring wells is avoided by leaving one clear well between those containing samples. Small corrections to data can be made if required by measuring the volume lost to these neighbouring, empty wells after centrifugation. However, this should not exceed 5% of the volume added.
[e] The force required to achieve contact will depend on the cell type. The value given here is sufficient to bring > 90% of chick embryo neural retina cells in contact with a monolayer of the same cell type. This should be checked for each cell type in preliminary experiments by removing the cell suspension after this first spin (at different speeds) and counting the labelled cells in the medium of each well.
[f] The speed of this spin to remove probe cells from the monolayer will depend on the adhesion molecules under study. Speeds equivalent to generating between 1 and 2000 g should be used.

Probe cells are brought into contact with the monolayers by centrifugation and, either during this spin or during incubation after this step, the nature of the adhesions formed can be investigated with respect to such parameters as temperature, the presence of metabolic inhibitors and the time for which the two cell populations are in contact. The plates are then sealed, inverted, and centrifuged once more to dislodge the probe cells. Weak adhesions that have formed require only mild centrifugal forces to break them. The bottoms of the wells can then be cut and the proportion of adherent cells determined by scintillation counting. A graphical representation of the percentage of cells bound under any particular set of conditions is probably suitable for most applications although calculations of the relative centrifugal force needed to remove populations of single cells from a cell monolayer are also possible (23). This, like the collision efficiency coefficient (α) described above, may provide a convenient index for quantitatively comparing different treatments or different cell lines. However, as with measurements of α, the biological relevance of the absolute value is questionable.

3.2 Measurements of specific adhesion

Specific adhesive mechanisms between two cell types are believed to play a fundamental role in embryogenesis and tissue organization, and such adhesive selectivity can be assayed with respect to individual molecules or to functional mechanisms. The involvement of such molecules in adhesive interactions between cells is often tested by the ability of specific antibodies to block the ability of surface CAMs to mediate intercellular adhesions. Typically, adhesion assays are performed according to procedures such as those described above with the introduction of peptides or specific antibodies into the assay to assess their ability to prevent the formation of multicellular aggregates (by binding to and thereby functionally inactivating CAMs). This type of assay has been particularly successful in elucidating the molecular identity of a large number of neural CAMs including N-CAM (28), N-cadherin (29) and L1 (30). In such perturbation assays it is essential to use Fab' fragments prepared from the antibodies. This is because

- the use of smaller molecules reduces the possibility that any observed inhibition of cell adhesion is due to non-specific steric hindrance at the cell surface, and
- being monovalent, Fab' fragments are capable of blocking inhibition following binding rather than inducing agglutination.

The production of Fab' fragments from whole IgG, together with their use in adhesion assays is described in *Protocol 6*.

Protocol 6. Testing IgG for an anti-CAM titre

Preparation of Fab' fragments[a]

1. Dissolve 10 mg IgG in 1.0 ml 20 mM NaH_2PO_4/20 mM cysteine/10 mM EDTA (pH 7.0)—buffer A.
2. Add 0.5 ml agarose-immobilized papain to this solution (Pierce Laboratories) and incubate it in a shaking water bath at 37°C for 5 h.
3. Add 3.0 ml 10 mM Tris–HCl (pH 7.5) and mix well.
4. Add iodoacetamide to a final concentration of 75 mM.
5. Spin the solution for 5 min at 500 g at room temperature (RT).

6. Transfer the supernatant to 5 ml agarose-immobilized protein A column, equilibrated with 10 mM Tris–HCl (pH 7.5). The pellet will contain some undigested IgG. To increase the recovery of Fab' fragments, wash and resuspend the pellet in 1.0 ml buffer A and repeat steps 2–6.
7. Elute the Fab' fragments with 15 ml 10 mM Tris–HCl (pH 7.5). If Fc fragments are also desired, these can be removed by washing the gel with 0.1 M glycine–HCl (pH 2.8). The Fc fraction should be neutralized immediately with solid Tris.
8. Concentrate the antibody by microfiltration or ammonium sulfate precipitation.
9. Dilute Fab' fragments to a final concentration of 2 mg/ml in MEM.[b]

Incubation of cells with Fab' fragments

10. Follow *Protocol 4* to step 6 but bring the cells to a concentration of 1×10^8 cells/ml.
11. Remove 0.05 ml of the cell suspension and add 0.05–0.5 ml Fab' solution.[c]
12. Incubate the cells with gentle swirling at 4°C for 20 min and dilute them to a final volume of 2.0 ml with ice-cold MEM.
13. Pass the cells twice through a 19-gauge syring needle to break up loose aggregates, and filter through a Nitex membrane.
14. Count the cells and introduce them into an appropriate adhesion assay.

[a] Prior to large-scale preparation of Fab' fragments, preliminary tests should be conducted to confirm that the amount of papain used and the incubation time are adequate.
[b] For high affinity antisera, the final Fab' concentration can be reduced.
[c] The specificity of the interaction can be tested by incubating the cells with an equivalent volume of the Fab' fragment solution which has previously been incubated for 20 min at 25°C with the fraction (usually either whole cells, membrane extracts, or purified proteins) used to immunize the animals.

Using the above strategy, individual molecules can be assessed for their ability to mediate intercellular adhesion. Using recombinant cDNA techniques it has recently been possible to confirm what was long suspected, namely that some neural CAMs are responsible for conferring adhesive specificity upon cells such that they sort out from each other when artificially mixed (31, 32). A standard technique for such studies is to transfect a population of suitable parental cells with cDNAs encoding the CAM of interest and then establish a clonal cell line. Qualitative and quantitative expression of the CAM can be determined and, if the chosen parental cell line does not normally express the CAM (as is usually the case), this provides a good negative control for many experiments. In such cases, a simple way for confirming the identity, and thereby distribution, of individual cells within an artificially-formed aggregate which may be composed of parental and transfected cells is to recognize them with relevant anti-CAM antibodies. However, if specific antibodies are not available then labelling one of the cell populations prior to aggregation is necessary. In addition to providing information on cell sorting, these approaches can also be used to test hypotheses as to whether individual CAMs are operating by either homophilic or heterophilic mechanisms. Indeed, using transfection and pre-labelling studies, I have shown that the major myelin protein Po (in common with many neural CAMs which are also members of the immunoglobulin superfamily) acts as a CAM which can operate by homophilic-binding mechanisms (16).

Protocol 7. Co-aggregation of differentially-labelled cell populations

1. Proceed from steps 1 to 3 of *Protocol 4* for the two cell populations under study.
2. Resuspend both suspensions in 10 ml ice-cold HBSS. To one suspension add 10 μl acetone and to the other add 10 μl of stock 5- (and 6-) carboxyfluorescein diacetate succinimyl ester (CFSE).[a]
3. Incubate both cell suspensions for 15 min at 37°C.
4. Centrifuge at 400 g for 10 min at 4°C.
5. Resuspend the cells in 15 ml HBSS at 4°C.
6. Repeat step 4.
7. Resuspend in either 2.0 ml MEM (for short-term assays) or full medium (either defined or serum-supplemented) for prolonged assays (see below) and gently pass the cells twice through a 21-gauge syringe needle.
8. Bring both suspensions to a concentration of 0.5×10^6 cells/ml[b] and filter them through a Nitex membrane.
9. Mix the suspensions and introduce into a suitable adhesion assay (rotational, Couette viscometry, etc.).
10. Remove either 20-μl samples after 5–30 min or 100-μl samples after 24 h.[c]
11. Transfer the cells to poly-L-lysine coated slides and allow them to attach for 5 min at room temperature.
12. Rinse in 0.1 M sodium cacodylate buffer (pH 7.2).
13. Fix the cells in 4% paraformaldehyde in cacodylate buffer for 30 min at room temperature.
14. View the aggregates by epifluorescence microscopy.

[a] CFSE (Molecular Probes, Oregon) is kept as a 10 mM stock solution in acetone.
[b] As in all adhesion assays and their variations, the exact numbers of cells to be used must be determined empirically in pilot experiments.
[c] Two adhesive phenomena can be measured by this assay: if cells are only allowed to aggregate for a brief period then initial specificities can be measured; if they are left for longer, then the ability of the cells to sort out is assayed (see text).

A simple method for identifying two different cell types in aggregates is described in *Protocol 7*. Analysis of the short-term assay described is based on the observation that if two cell suspensions are allowed to aggregate, then non-specific adhesion is characterized by a binomial distribution of the two cell types within the resulting aggregates (33). Deviations from such distributions, predicted on the basis of random aggregation, can then be measured and used as an indication of specific adhesive mechanisms operating between the two cell populations (8, 17, 27, 33). Diagrammatic representations of distributions which reflect no adhesive specificity (random aggregation), partial and high specific adhesions between cells are shown in *Figure 3*. I have used this

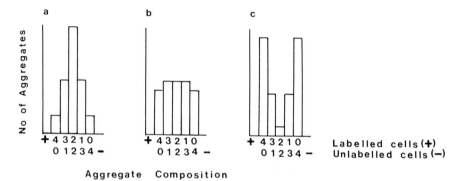

Figure 3. Representations of possible frequency distributions of 4-cell aggregates, assuming an equal ratio of labelled (+) and unlabelled (−) cells. In **a**, the histogram representing non-specific adhesion is shown and is characterized by a binomial (1:3:6:3:1) distribution. In **b**, partial adhesive specificity between the two cell populations results in an increase in the number of homotypic aggregates and a corresponding decrease in the proportion of those of mixed composition. In **c**, marked specific adhesion between the labelled and unlabelled cells is reflected by a bimodal distribution essentially consisting of two distinct populations of homotypic aggregates.

approach to demonstrate that, within the limits of the adhesion assay used, no adhesive specificity exists between 3T3 cells transfected with cDNAs encoding either lipid-linked or transmembrane isoforms of N-CAM, although transfected 3T3 cells expressing a potentially secreted N-CAM variant did not aggregate randomly with those expressing these other N-CAM variants (5).

Analysis of such assays must be conducted on small aggregates, since it is essential to be able to determine the total number of labelled and unlabelled cells within them. For this reason, samples must

- be confirmed to consist only of single cells at the start of the assay, and
- only be allowed to aggregate for a brief period.

Furthermore, for ease of computation, it is convenient to select only one size class of aggregate (for example, only those consisting of four cells) for analysis.

After aggregation, approximately 200 particles (cells and aggregates) falling on randomly-drawn orthogonal diameters should be counted and the proportion of stained (m) and unstained (n) cells counted. These values should be within 5% of the original proportions at the start of the assay to ensure that preferential loss of one cell type due, for example, to lysis, has not occurred during the course of the assay resulting in bias during subsequent evaluation. A particular size class is then selected and the number of labelled and unlabelled cells within 50–100 aggregates is recorded. The deviation of the observed distribution from one which assumes no adhesive selectivity between the two cell populations can then be measured by a chi-square test where N is the number of aggregates of the particular size class, and s^2 and σ^2 are the variances of the observed and calculated binomial distributions respectively. A series approximation can then be used to calculate the P values for s^2 with v ($v = N - 1$) degrees of freedom. Thus, in a sample (N) of 4-cell aggregates consisting of A and B cell types, the mean content of A cells in the aggregates ($\overline{A}$) can be determined, and the variance (s^2) is given by:

$$s^2 = [\Sigma(A - \overline{A})^2]/N - 1.$$

From analysis of the aggregates, the relative frequencies of A (m) and B (n) cells can be determined and thus the binomial distribution which assumes random adhesion between the cells in the 4-cell aggregate class is described by:

$$P(A) = N(4!/[A!(4 - A)!])(m^A)(n^{1-A})$$

and its variance (σ^2) is 4 mn.

The above analysis measures the initial ability of a cell population to aggregate randomly with a different cell type or for the two groups of cells to preferentially segregate due to different adhesive surface properties, for example due to the presence of different CAMs on their surface. An alternative approach to adhesive specificity concerns the ability of cells to sort out from one another during prolonged periods of aggregation. In these cases, any specificity cannot easily be quantitated since the aggregates become too large to determine the numbers of cells within them. However, if cells in such assays selectively adhere to their own type, then this is readily observed by the presence of labelled and unlabelled domains within the aggregates. A variation of this assay has been used by Nose et al. (31), who rapidly induced aggregation of two cell populations by the addition of 0.1 mg/ml soybean agglutinin to the aggregation medium. This results in an initial random mixture of cells within the aggregates, and the ability of the cells to subsequently segregate during culture of the whole aggregates is then assessed. In this way, for example, it was found that transfected L cells expressing related cadherins (either E-cadherin or P-cadherin), sort out from each other (31), confirming earlier studies that cells of different origin which normally express different cadherin molecules segregate from each other *in vitro* (34).

4. Neurite outgrowth

4.1 Basic principles

Neural pathways are established with great precision in the development of the nervous system. It is generally believed that this is achieved by the motile growth cones of neurites interacting with guidance cues in their environment (for reviews, see refs 35, 36). Such extrinsic cues can conveniently be divided into two classes: those derived from components of the extracellular matrix and those which are associated with the membranes of both neural and non-neural cells. These divisions can be regarded as similar to the distinctions between SAMs and CAMs described in Section 1. However, it is important to note that, as described above, there are certain overlaps within these classifications and the allocation of a particular molecule to one or other class is sometimes arbitrary. Another noteworthy point is that these molecules, whether involved in intercellular adhesion or in neurite outgrowth, should not be regarded as a molecular 'glue', acting independently of other cellular processes. For example, in addition to the homophilic affinity of the extracellular domains of many neural CAMs, their cytoplasmic regions are also often essential for normal function (32, 37). These intracellular interactions may involve signal transduction mechanisms, direct involvement with the cellular cytoskeleton, or both. Finally, although some neural CAMs (such as N-CAM, L1 and N-cadherin) can function both to mediate intercellular adhesion and promote neurite outgrowth, it should not be assumed that the same mechanisms are operating in both cases. Similarly, it should not be assumed that neurite outgrowth is simply a consequence of neural cells encountering an adhesive substratum. Indeed, some proteoglycans and their component glycosaminoglycan chains have been shown to inhibit neurite outgrowth under some circumstances (38).

A great variety of *in vitro* assays have been developed to measure neurite outgrowth, but nearly all involve culturing neuronal cells upon a suitable substratum (as with cell-substratum adhesion assays, these are usually represented by cellular, purified or semi-purified protein) and then measuring neuritic extension by morphological or biochemical criteria. In addition, several principles and applications described in the preceding sections on intercellular adhesion apply to studies of neurite outgrowth. For example, the use of the Fab' fragments in antibody-perturbation assays and cDNA transfection strategies have frequently been used to elucidate the identity and roles of molecules involved in this process (for examples, see refs 39–41) and it should be remembered that, again in common with intercellular adhesion, more than one class of molecules are involved; neurite outgrowth is a dynamic process which involves the presence of different CAMs in addition to SAMs, growth factors and other molecular families. Furthermore, the expression of many (if not all) neurite outgrowth-promoting molecules are under temporal as well as tissue-specific control. For this reason, the developmental age of any neurons under study must also be considered with respect to individual CAM expression *in vivo*.

4.2 Neurite outgrowth assays

In this section, three different assays are described which have been used to study the requirements for neuronal differentiation and some of the molecules involved. I have maintained the theme of the first section in that emphasis is given to assays based on cell–cell interactions rather than to assays involving measurements of neurite outgrowth on non-cellular coated substrata. However, it will easily be seen how the assays described can be modified for such applications. It is also acknowledged that, even within these criteria, many varied *in vitro* assays of neurite outgrowth assays have been omitted; the intention being to describe simple assays from which much valuable information can be rapidly gained.

4.2.1 Neurite outgrowth on cultured cell monolayers

The role of individual cell types (or of cells expressing specific, chosen molecules) in mediating neuronal extension can conveniently be assayed by testing the ability of such cells to act as a permissive substratum for a particular neuronal population and under defined conditions. The use of 8-chamber tissue-culture slides (for example, those supplied by Lab-Tek®) allow for multiple experiments to be conducted simultaneously on the same parent cultures and in an identical microenvironment. Typically, a monolayer is established within the chambers of the slides (which may first need to be coated with collagen, laminin or other suitable attachment factor according to the cell type; see also Chapter 3) and, after a suitable period, neuronal cells are introduced to the medium bathing the monolayer. These cells are then allowed to adhere and after an appropriate incubation time neurite extension can be identified by indirect immunological methods. A range of appropriate primary antibodies are commercially available, but convenient ones include antineurofilament antibodies (Sigma provide a panel of antibodies which recognize mammalian neurofilament polypeptides of different molecular weights and which do not cross-react with other intermediate filament proteins) and the monoclonal antibody A2B5 which recognizes gangliosides on the surface of neurons and type II astrocytes (42, 43). Hybridomas secreting A2B5 are available from the European Collections of Animal Cell Cultures held at Porton Down. Direct morphological measurements can then be made using such indices as the total length of neuritic extensions, the length of the longest single neurite and the degree of branching. This is often performed simply using camera lucida systems, but several studies have also utilized video image analysis for quantitation of neurite outgrowth (44–48). Several simpler methods of neurite identification are also frequently employed, such as observation of neurons stained with Coomassie blue (0.1% in 10% acetic acid, 25% propanol) and counting the proportion which extend neurites (of any length) or the proportion of neurites greater than a defined length. These forms of analysis are prone to miss small (but potentially significant) changes in neurite length, since fine neuritic processes are sometimes undetected by Coomassie-blue-staining, as well as more fundamental effects such as those which may affect branching but not elongation, or vice versa.

Protocol 8. Measurements of neurite outgrowth from DRG neurons on 3T3 monolayers

Establishment of cell monolayers

1. Grow 3T3 fibroblasts routinely in media such as Dulbecco's MEM supplemented with 10% fetal calf serum.
2. Coat the wells of 8-chamber slides with collagen[a] before use and keep at 4°C.
3. Dissociate cells using methods such as those described in *Protocols 1* and *2*. Since the cells will be allowed to recover, temporary inactivation, or removal of surface CAMs is unimportant.
4. After washing, resuspend the cells at a final concentration of 1.0×10^6 in full medium and dispense 100 µl of the suspension into each chamber of the slides.
5. Maintain 3T3 cultures for 48 h before addition of neurons.

Preparation of DRG neurons

6. Remove approximately 100 DRG from 1-day-old mice and incubate in 2.0 ml DMEM/10 mM Hepes/0.2% trypsin/0.03% collagenase/0.01% DNase I for 45 min at 37°C.
7. Inactivate the trypsin by the addition of 1.0 ml DMEM/10% FCS and dissociate the ganglia by gently triturating through a 21-gauge syringe needle.
8. Spin the suspension at 4°C for 5 min at 80 g.
9. Resuspend the pellet in 200 µl DMEM/10 mM Hepes/10% horse serum (culture medium).
10. Layer the suspension on a 1.0 ml cushion of 35% Percoll and spin at 200 g for 15 min at 4°C.
11. Wash the cells in 5.0 ml culture medium and spin as in step 8.
12. Resuspend the pellet in 1.0 ml culture medium, triturate through a 21-gauge syringe needle and pass the cells through a sterile 20 µm Nitex membrane.
13. Dilute the single cell suspension to a concentration of 10^4 cells/ml in culture medium supplemented with β-nerve growth factor (NGF) at a final concentration of 100 ng/ml.
14. Allow DRG neurons to attach to cell monolayers and extend neurites for 1–7 days.

Protocol 8. *Continued*

Determination of neurite outgrowth

15. Fix cells with 4% paraformaldehyde in culture medium for 30 min at RT.
16. Permeabilize cells and block non-specific binding with 200 μl PBS/1% FCS/0.05% saponin for 15 min at RT.[b]
17. Wash the cells three times in PBS, and incubate them for 1 h at room temperature with 50 μl of appropriate primary antibody (see text) diluted in PBS/0.1% BSA.
18. Wash the cells three times in PBS/0.1% BSA.
19. Incubate the cells for 1 h at RT with 50 μl appropriate rhodamine isothiocyanate-coupled secondary antibody diluted in PBS/0.1% BSA.
20. Wash the cells well in PBS, peel the chambers from the slides, mount them in aqueous mounting media, and visualize neurites by fluorescence microscopy for morphological measurements.

[a] Purified rat-tail collagen is available as Vitrogen® from Collagen Corporation (Palo Alto, California).
[b] If the primary antibody is directed against an extracellular epitope of a surface molecule, permeabilization is not required and therefore saponin should be omitted.

A method for quantitatively measuring neurite growth is given in *Protocol 8* where the example given measures the ability of 3T3 fibroblasts to promote the outgrowth of neurites from dorsal root ganglion (DRG) neurons. This type of assay has been used previously to assess the abilities of cell substrata such as L cells, astrocytes and Schwann cells (among others) to promote neurite outgrowth from these neurons (49–51). However, the basic assay described should be considered a template for others in which peptides, biochemical perturbants, antibody Fab' fragments or cells which have been manipulated by recombinant cDNA techniques can be introduced. Such applications are clearly a requirement for dealing with individual molecules known or suspected of modulating neurite outgrowth.

An alternative strategy which also allows for measurements to be made on multiple parallel cultures involves seeding the cells which will constitute the monolayer into the wells of 96-well microtitre plates. After the monolayer has become established, neuronal cells are introduced as before. Once more, neuronal cells and their associated neurites can be identified by appropriate antibodies. However, this time the signal is detected by ELISA. Alternatively, the same ELISA technique can be used following incubation of a primary antibody directed against specific molecules either known to be involved, or implicated, in neuronal development. For example, this form of analysis has been successfully (on non-cellular substrata) to investigate the expression of L1, N-CAM, and Thy-1 in PC 12 cells (52) and the monoganglioside GM1 in neurite outgrowth assays using chick DRG neurons (53).

Analyses based directly on morphological measurements and those utilizing ELISA both have advantages and disadvantages. Perhaps the most obvious appeal of the latter is the facility to process a much larger number of replicate or related experiments, and to achieve this very rapidly. The assay is also easy to quantitate and provides highly reproducible data. The major disadvantage of this assay is that only one index of neurite outgrowth, that related to total neuritic extension, can be evaluated. Furthermore, a large component of the signal generated will usually be a consequence of antibody binding to the cell body of the neurons rather than to the neurites themselves. However, this should not prove a major problem if appropriate controls are used and there are no significant changes which could affect antibody binding within the bodies of the cells which are actively extending neurites.

Measurements using image analysis have obvious advantages in that several morphological parameters of neurite outgrowth can be made, as mentioned above. In addition, the health of the cultures can be monitored while taking measurements and errors due to contribution in the ELISA from dead or dying cells can be obviated. Unfortunately, although multiple experiments can also be performed rapidly for this form of subsequent analysis, measurements of the neurites is a much more laborious process, even with complex image processing systems. Another factor that should be considered is that although signals from the cell bodies can be avoided by this analysis, it is obviously impossible to account for any differences that may occur in the calibre of the neurites by measurements of length alone. Such increases may reflect a significant increase or decrease in the levels of neurite-associated molecules and would usually be detectable using ELISA methods. However, several image analysis packages allow for quantitative immunofluorescence measurements to be performed which circumvents this problem. Additionally, if the primary antibodies are directed against specific neural molecules of interest (as opposed to simply being markers of neuronal processes) then the immunofluorescent signal can be used to quantitate changes in the level of expression of such molecules under different experimental conditions. This can then be correlated directly with morphological data obtained from the same neuronal population (46). Recently, an assay combining both ELISA and quantitative video image analysis has been described in studies of GAP-43 expression and neurite outgrowth in PC 12 cells (48). This essentially involves making morphological measurements on Coomassie-blue-stained cultures following standard ELISA procedures and is described in *Protocol 9*.

Protocol 9. Combined analysis of neurite outgrowth in PC 12 cells by ELISA and image analysis[a]

Preparation of PC 12 cultures

1. Coat the wells of 96-well microtitre plates with 80 μl poly-L-lysine (50 μg/ml) or test substratum for 1 h.

Protocol 9. *Continued*

2. Grow, dissociate, and wash PC 12 cells in Sato supplemented with NGF (50 ng/ml) using routine methods (18).
3. Dilute the cells to a concentration of 1.25×10^5 cells/ml and aliquot 80 μl into each well.
4. Allow the cells to adhere for 1.5 h at RT, and then add NGF to a final concentration of 50 ng/ml.
5. Culture the cells for 0–72 h.
6. Fix the cells by adding 1 vol. 4% paraformaldehyde in DMEM for 1 h at RT, then carefully aspirate the medium and replace it with 4% paraformaldehyde in DMEM and leave for a further 1 h at RT.
7. Wash the cells well in PBS and permeabilize them with methanol for 20 min at $-20°C$.[b]

ELISA protocol

8. Rinse the cultures in PBS (3 times, 5 min each).
9. Block non-specific binding with PBS/10% horse serum.
10. Incubate the plates overnight with 50 μl appropriate primary antibody in PBS/10% horse serum/0.2% Triton X-100 at 4°C.[c]
11. Wash the cultures three times with PBS (pH 8.6).
12. Incubate the cultures with 50 μl appropriate secondary antibody $F(ab)_2$ fragments coupled to horseradish peroxidase in PBS (pH 8.6)/10% horse serum for 1 h at 37°C.
13. Wash the cells three times with PBS (pH 8.6) and twice with distilled water.
14. Incubate the plates with 50 μl *o*-phenylenediamine (200 mg/ml) in 0.1 M citrate buffer/0.2 M Na_2HPO_4/0.02% H_2O_2 for 20 min at 37°C.
15. Stop the reaction by the addition of 50 μl 4.5 M H_2SO_4.
16. Determine the optical densities of wells at 492 nm in an ELISA reader.[d]
17. Rinse the cultures three times in distilled water and stain them for 4 min with 0.1% Coomassie brilliant blue R250 in 10% acetic acid/25% propanol[e] and use directly for image analysis using bright-field microscopy.

[a] After ref. 48.
[b] See footnote b in *Protocol 8*.
[c] Ref. 48 recommends incubating overnight at 37°C. However, this may produce unacceptably high background with some antibodies and either 45 min at 37°C, or overnight at 4°C may yield better results.
[d] ELISA readers for direct assay of peroxidase activity in microtitre wells are available from several suppliers (for example, Bio-Rad or Dynatech) and interfaces with computers are also available.
[e] Ref. 48 gives opposite concentrations for acetic acid and propanol, but those given in the procotol are more commonly used.

Protocol 10. Neurite outgrowth on cryostat sections of muscle

1. Remove suitable muscle (such as diaphragm) from adult rats, immerse in Tissue-Tek (Miles Laboratories), and freeze rapidly in isopentane cooled by liquid nitrogen and store at −70°C.
2. Cut 5-μm sections at −20°C and collect on sterile 15-mm glass coverslips which have been coated overnight with poly-L-lysine (1 mg/ml in borate buffer, pH 9.0). Allow sections to air-dry on to coverslips in a sterile environment.
3. Place the coverslips bearing tissue sections in the wells of 35-mm tissue-culture dishes and wash them well in culture medium (see below).
4. Prepare single neural cell suspensions and dilute to 2×10^5 cells/ml in the appropriate medium. For example, if DRG neurons are prepared, use the culture medium described in *Protocol 8*. If PC 12 cells are used, incubate in Sato media as described in the footnote to *Protocol 3*. Supplement the media with 50 ng/ml NGF.[a]
5. Dispense 200 μl of the cell suspension on to each coverslip and culture for 24 h.
6. Add fresh medium and culture the cells for a further 1–6 days to allow for neurite outgrowth.
7. Wash the cells in DMEM, fix in 4% paraformaldehyde in DMEM for 1 h at RT and either wash in PBS and stain with 0.1% Coomassie blue for 4 min (step 17, *Protocol 9*) or process for immunological detection as described in steps 16–20, *Protocol 8*).
8. Wash the cells well in PBS, mount in aqueous media and assess neurite outgrowth by either bright field or fluorescence microscopy, as appropriate.

[a] Irrespective of the medium used, it should be supplemented with 20 U/ml gentamicin.

4.2.2 Neurite outgrowth on cryostat sections

As mentioned above, several molecules have been suggested to be candidates for mediating neurite outgrowth and axonal guidance. However, most of the assays (particularly those involving antibody-perturbation) which have identified such molecules, have used cultured cells. Although such *in vitro* studies have yielded much valuable information, it must always be considered that they can only, at best, approximate the molecular events of neuronal differentiation *in vivo*. Unfortunately, studies of the expression and function of these molecules in whole animals are often difficult to interpret.

One method that has been used as a compromise to the problems presented by both *in vivo* and *in vitro* studies of neurite outgrowth originates from the work of Stamper and Woodruff (54, 55) who developed a 'cryoculture' technique. This involves using cryostat sections as a substratum for cultured cells, and has been used in studies of neurite outgrowth and axonal guidance of PC 12 cells, DRG neurons, and sympathetic neurons on cryostat sections derived from various tissues (56–58).

This form of assay also has obvious limitations, not the least being that a frozen section could be regarded as no more representing a physiological substratum than that presented by a population of cultured cells. Perhaps the main advantages of the 'cryoculture' assays are that a variety of tissue structures are available for neurons to interact with, and it is easy to provide a choice of different substrata in the same assay by simply seeding a neuronal population on to coverslips which contain tissues of different origin. For example, the preference for DRG neurons to extend neurites which follow the basal lamina (or structures which cannot be resolved from it by light microscopy) in frozen muscle sections was determined in this way (57). The same study made use of the defined motor end-plate regions which exist in the diaphragm to monitor the behaviour of neurons differentiating over end-plate-rich and end-plate-free domains of this muscle.

As with the other assays of neurite outgrowth described here, neurites can be identified by different detection systems at the end of the experiment. Both Coomassie-blue-staining and immunological procedures have been used, but the main criterion for choice (unless the expression of a single molecule is desired for characterizing neurites) should be the facility with which the neurons can be distinguished from the underlying cryostat section. In practice, this does not pose a great problem, and indeed such distinction is often easier than dealing with a substratum which is also composed of cultured cells. Finally, if the coverslips used to collect the sections are sterile, handling of the sections is kept to a minimum, and gentamicin or other suitable antibiotics are kept in the culture medium, then contamination of the cultures is relatively rare. If contamination problems cannot easily be resolved, sterile measures can be augmented by UV-irradiating the sections prior to adding the neurons.

5. Future prospects

As described earlier in this chapter, studies of intercellular adhesion and neurite outgrowth have progressed from a phenomenological basis to analyses of the individual molecules involved. Much of the increase in knowledge is due to the development and application of immunological and molecular biological procedures. These approaches, particularly the latter, are likely to continue to contribute much information to the identification of new CAMs and molecules capable of promoting neurite outgrowth. In addition to the identification of novel molecules, much interest is currently being given to analyses relating to control of CAM expression and of functional domains within them. These studies often involve cDNA transfection using either chimaeric, deleted or site-mutated constructs.

A complete understanding of the regulation of CAM expression may prove to be extremely difficult to determine. This is partly because different isoforms of the same molecule can represent great heterogeneity with respect to extracellular or cytoplasmic domains, carbohydrate modifications or phosphorylation status. Each form may be under developmental stage- or tissue-specific control with its own set of factors required for expression and normal function. For example, it has been reported that at least 27 alternatively spliced mRNAs from the single N-CAM gene are expressed during development of the rat heart (59). Elucidation of all the splicing factors involved, and their control, will clearly take considerable effort.

In addition to functional studies being conducted on cells following the introduction of constitutively expressed genes, an alternative strategy is to inactivate specific genes and then analyse the ability of cells to adhere or extend neurites. This can be performed in different ways, but perhaps the commonest *in vitro* technique is by the transfection of antisense constructs of specific cDNAs to suppress expression. This approach has been used to demonstrate a role for the glial fibrillary acidic protein in promoting the extension of astrocytic processes in the presence of neurons (60). Other means that are currently being used to selectively inhibit the expression of genes involved in adhesion or neurite outgrowth include incubating cells with antisense oligonucleotides which are complementary to the 5' sequence of the gene under study. In this way, it has been shown that inhibition of expression of the microtubule-associated protein tau results in a failure of cerebellar neurons to extend long, axon-like neurites while having no apparent effect on the production of shorter neuronal processes (61).

These studies, involving the activation or repression of gene expression are likely to continue to contribute greatly to the studies of the roles of individual molecules in adhesion and neuronal differentiation *in vitro*, and to complement analysis on whole embryos or animals by such techniques as homologous recombination and the production of transgenic animals. Finally, the dynamic nature and complexity of the cellular processes described in this section will necessitate the investigation of potential co-operation (positive or negative) between different molecular species involved in cell adhesion and neurite outgrowth.

References

1. Edelman, G. M. (1986). *Ann. Rev. Cell Biol.*, **2**, 81.
2. Takeichi, M. (1988). *Development*, **102**, 639.
3. Volk, T. and Geiger, B. (1986). *J. Cell Biol.*, **103**, 1441.
4. Gower, H. J., Barton, C. H., Elsom, V. L., Thompson, J., Moore, S. E., Dickson, J., and Walsh, F. S. (1988). *Cell*, **55**, 955.
5. Pizzey, J. A., Rowett, L. H., Barton, C. H., Dickson, J. G., and Walsh, F. S. (1989). *J. Cell Biol.*, **109**, 3465.

6. Weiss, L. (1961). *Exp. Cell Res.*, **8**, 141.
7. Magnani, J. L., Thomas, W. A., and Steinberg, M. S. (1981). *Devel. Biol.*, **81**, 96.
8. Gibraltar, D. and Turner, D. C. (1985). *Devel. Biol.*, **112**, 292.
9. Knudsen, K. A., Myers, L., and McElwee, S. A. (1990). *Exp. Cell Res.*, **188**, 175.
10. Takeichi, M., Atsumi, T., Yoshida, C., Uno, K., and Okada, T. S. (1981). *Devel. Biol.*, **87**, 340.
11. Jessell, T. M. (1988). *Neuron*, **1**, 3.
12. Takeichi, M. (1987). *Trends Genet.*, **3**, 213.
13. Williams, A. F. and Barclay, A. N. (1988). *Ann. Rev. Immunol.*, **6**, 381.
14. Cunningham, B. A., Hemperly, J. J., Murray, B. A., Prediger, E. A., Brackenbury, R., and Edelman, G. M. (1987). *Science*, **236**, 799.
15. Moos, M., Tacke, R., Scherer, H., Teplow, D., Fruh, K., and Schachner, M. (1988). *Nature*, **334**, 701.
16. Filbin, M. T., Walsh, F. S., Trapp, B. D., Pizzey, J. A., and Tennekoon, G. I. (1990). *Nature*, **344**, 871.
17. Pizzey, J. A., Jones, G. E., and Walsh, F. S. (1988). *J. Cell Biol.*, **107**, 2307.
18. Bottenstein, J. E. (1985). In *Cell culture in the neurosciences* (ed. J. E. Bottenstein and G. Sato). Plenum Press, London.
19. Cole, G. J., Loewy, A., and Glaser, L. (1986). *Nature*, **320**, 445.
20. Curtis, A. and Lackie, J. (1991). In *Practical aspects of cell adhesion* (ed. A. S. G. Curtis and J. M. Lackie), pp. 1–22. Wiley, Chichester.
21. Garrod, D. R. and Born, G. V. R. (1971). *J. Cell Sci.*, **8**, 751.
22. Skehan, P. (1975). *J. Membr. Biol.*, **24**, 87.
23. McClay, D. R., Wessel, G. M., and Marchase, R. B. (1981). *Proc. Natl. Acad. Sci. USA*, **78**, 4975.
24. Garrod, D. R. and Nicol, A. (1981). *Biological reviews of the Cambridge Philosophical Society*, **56**, 199.
25. Curtis, A. S. G. (1969). *J. Emb. Exp. Morph.*, **79**, 305.
26. Jones, G. (1991). In *Practical aspects of cell adhesion* (ed. A. S. G. Curtis and J. M. Lackie), pp. 23–39. Wiley, Chichester.
27. Pizzey, J. A. and Jones, G. E. (1985). *J. Neurol. Sci.*, **69**, 207.
28. Thiery, J.-P., Brackenbury, R., Rutishauser, U., and Edelman, G. M. (1977). *J. Biol. Chem.*, **252**, 6841.
29. Yoshida, C. and Takeichi, M. (1982). *Cell*, **28**, 217.
30. Grumet, M. and Edelman, G. M. (1984). *J. Cell Biol.*, **98**, 1746.
31. Nose, A., Nagafuchi, A., and Takeichi, M. (1988). *Cell*, **54**, 993.
32. Jaffe, S. H., Freidlander, D. R., Matsuzaki, F., Crossin, K. L., Cunningham, B. A., and Edelman, G. M. (1990). *Proc. Natl. Acad. Sci. USA*, **87**, 3589.
33. Sieber, F. and Roseman, S. (1981). *J. Cell Biol.*, **90**, 55.
34. Nose, A. and Takeichi, M. (1986). *J. Cell Biol.*, **103**, 2649.
35. Doherty, P. and Walsh, F. S. (1989). *Current opinions in cell biology*, **1**, 1102.
36. Sanes, J. R. (1989). *Ann. Rev. Neurosci.*, **12**, 491.
37. Nagafuchi, A. and Takeichi, M. (1988). *EMBO J.*, **7**, 3679.
38. Akeson, R. and Warren, S. L. (1986). *Exp. Cell Res.*, **162**, 347.
39. Neugebauer, K. M., Tomaselli, K. J., Lilien, J., and Reichardt, L. F. (1988). *J. Cell Biol.*, **107**, 1177.
40. Matsunaga, M., Hatta, K., Nagafuchi, A., and Takeichi, M. (1988). *Nature*, **334**, 62.

41. Doherty, P., Rowett, L. H., Moore, S. E., Mann, D. A., and Walsh, F. S. (1991). *Neuron,* **6,** 247.
42. Eisenbarth, G. S., Walsh, F. S., and Nirenberg, M. (1979). *Proc. Natl. Acad. Sci. USA,* **76,** 4913.
43. Raff, M. C., Abney, E. R., Cohen, J., Lindsay, R., and Noble, M. (1983). *J. Neurosci.,* **3,** 1289.
44. Gaver, A. and Schwatz, M. (1983). *J. Neurosci. Meth.,* **7,** 275.
45. Ford-Holevinski, T. S., Dahlberg, T. A., and Agranoff, B. W. (1986). *Brain Res.,* **368,** 339.
46. Doherty, P., Cohen, J., and Walsh, F. S. (1990). *Neuron,* **5,** 209.
47. Matsumoto, T., Oshima, K., Miyamoto, A., Sakurai, M., Goto, M., and Hayashi, S. (1990). *J. Neurosci. Meth.,* **31,** 153.
48. Jap Tjoen San, E. R. A., Schmidt-Michels, M. H., Spruijt, B. M., Oestreigher, A. B., Schotman, P., and Gispen, W. H. (1991). *J. Neurosci. Res.,* **29,** 149.
49. Seilheimer, B. and Schachner, M. (1988). *J. Cell Biol.,* **107,** 341.
50. Tomaselli, K. J., Neugebauer, K. M., Bixby, J. L., Lilien, J., and Reichardt, L. F. (1988). *Neuron,* **1,** 33.
51. Doherty, P., Barton, C. H., Dickson, G., Seaton, P., Rowett, L. H., Moore, S. E., *et al.* (1989). *J. Cell Biol.,* **109,** 789.
52. Mann, D. A., Doherty, P., and Walsh, F. S. (1989). *J. Neurosci.,* **53,** 1581.
53. Doherty, P. and Walsh, F. S. (1987). *J. Neurochem.,* **48,** 1237.
54. Stamper, H. B. and Woodruff, J. J. (1976). *J. Exp. Med.,* **144,** 828.
55. Woodruff, J. J., Clarke, L. M., and Chin, Y. H. (1987). *Ann. Rev. Immunol.,* **5,** 201.
56. Carbonetto, S., Evans, D., and Cochard, P. (1987). *J. Neurosci.,* **7,** 610.
57. Covault, J., Cunningham, J. M., and Sanes, J. R. (1987). *J. Cell Biol.,* **105,** 2479.
58. Sandrock, A. W. and Matthew, W. D. (1987). *Proc. Natl. Acad. Sci. USA,* **84,** 6934.
59. Reyes, A. A., Small, S. J., and Akeson, R. (1991). *Mol. Cell. Biol.,* **11,** 1654.
60. Weinstein, D. E., Shelanski, M. L., and Liem, R. K. H. (1991). *J. Cell Biol.,* **112,** 1205.
61. Caceres, A. and Kosik, K. S. (1990). *Nature,* **343,** 461.
62. Segal, D. M. (1988). In *Physical basis of cell–cell adhesion* (ed. P. Bongrand). CRC Press, Boca Raton, Florida.
63. Umbreit, J. and Roseman, S. (1975). *J. Biol. Chem.,* **250,** 9368.
64. Guarnaccia, S. P. and Schnaar, R. L. (1982). *J. Biol. Chem.,* **257,** 14288.

A1

A catalogue of neuronal properties expressed by cell lines

IAIN F. JAMES and JOHN N. WOOD

1. Introduction

The use of cell lines as research tools is so well established that it hardly needs to be justified. They have been central to studies of neuropeptide processing (see ref. 10), the discovery, characterization, and cloning of receptor subunits (32) and the regulation of ion-channel activity (172), as well as innumerable other studies. In this brief chapter we have catalogued and referenced the use of cell lines in studies of neuropeptides, neurotransmitter receptors, and voltage-gated ion channels. The already long reference list is by no means comprehensive, but we hope that it will provide a starting point for neuroscientists who may have a use for cell-based studies in their particular research area.

Table 1. Cell lines that express neuropeptides

Peptide	Cell line	Reference
Enkephalins	NG108-15	1, 6
	C6	2, 9
	SK-N-MC	3
	PC12	4
	SK-N-MCIXC	5
	N1E-115	7
	ROS 17/2.8	8
	ROS 25/1	8
Pro-opiomelanocortin	AtT-20	10
	HRE-H9	11
	SCLC	12
	NG108-15	6
	HMS-41/I	13
	HMS-78/2	13
	HMS-98/2	13
Dynorphin	R2C	14
	NG108-15	6

Table 1. (contd.)

Peptide	Cell line	Reference
Substance P	F11	15, 151
	ND5	15
	ND7	15
	ND11	15
	ND21	15
	ND C	15
	ND D	15
	AR42J	28
CGRP	UMR 106.01	16
	44-2C	17
	ND21	15
Calcitonin	CA-77	18
	HMS-41/I	13
	HMS-78/2	13
	HMS-98/2	13
CCK	WE4/2	19
	SK-N-MCIXC	5
	SN-N-MC	9, 20
	CA-77	21
Somatostatin	44-2C	22
	RIN	23
	HeLa	23
	ND21	15
Neuropeptide Y	PC12	24, 25, 26
	NS 20-Y	25
	LA-N-5	25
	CHP-234	25
	SMS-KCNR	25
	SH-SY5Y	25
	SMS-KCN	25
	BE(2)-M17	25
	SK-N-MCIXC	5
	NG108-15	26
	N18TG-2	26

Table 2. Cell lines that express peptide receptors

Receptor	Subtype	Cell line	Reference
Opioid	Mu	MCF 7	28
		SH-SY5Y	29
		SK-N-SH	35, 41
	Delta	NG108-15	30, 31, 32, 42
		P388dl	33
		NCB 20	34, 36, 42
		SK-N-SH	35, 41

Table 2. (contd.)

Receptor	Subtype	Cell line	Reference
		Y 79	41
		NHT C10	41
		IMR-32	41
		NMB	41
		F11	151
	Kappa	P388dl	33
	Unclassified	S20Y	37, 38
		NCI-H146	39
		PC12	44
Substance P	NK1	U373 MG	40
		AR 42J	45
CGRP		T47D	43
Calcitonin		T47D	46, 48
		BEN	46, 48, 50, 51
		BIN-67	47
		MCF 7	48
		LLC-PK1	49
CCK	CCK A	CHP212	52
	CCK B	JURKAT	53
Somatostatin		HGT-1	54, 55, 57
		AtT-20	56
		MIA PaCa-2	58
Neuropeptide Y		SK-N-MC	59, 60, 61
		SMS-MSN	62
		SMS-KAN	62
		CHP-243	62
		MC-IXC	62
		CHP-212	62
VIP		WE-68	63
		HT-29	64
		SH-SY5Y	65
		CL.16E	66
		SK-N-MC	59
		PC12	67
Bradykinin		F11	15, 68, 74, 151
		N1E-115	69
		NG108-15	70
		HSDM1C1	71, 74
		PC12	72
		NG115-401L	73
		ND7/23	15
Bombesin		Swiss 3T3	75, 79–82
		U-118	76
		AR42J	77, 78
		GH4C1	83

Table 3. Cell lines expressing neurotransmitter receptors

Receptor	Subtype	Cell line	Reference
Nicotinic		TE671	84, 88
		PC12	85
		IMR 32	86
		N1E 115	87
Muscarinic		M1-CHO	89
		SK-N-SH	90, 91, 95
	M3	C62B	92
		1321N1	93
		GH3	94
	M4	PC12	95
	M3	C6	95
	M2	IMR-32	95
	M2	Neuro-2A	96
		N1E-115	97
Alpha-1 Adrenergic		MDCK-D1	98
		RSMT-A5	99
		BC3H1	100
		DDT1	101
Alpha-2-Adrenergic		RINm5F	102
		C127	103
	alpha-2A	HT29	104
	alpha-2B	NG108-15	104
		SK-N-SH	90
		Y-79	105
		5H-SY5Y	105
	alpha-2C	OK	106
Beta-Adrenergic	Beta-1	SK-N-MC	107
	Beta-1, Beta-2	3T3-L1	108
		Y-79	109
		L6	110
		Cyc-S49	110
		NG108-15	111
		HIVE	112
		BFAE	112
		C6	112
		A431	113
	Beta-2	DDT1 MF-2	114
		L6	115
		3T3	116
		COLO 16	117
5-HT	Novel, 5HT1	NCB 20	118, 123
	5HT-1B	OK	119
	5HT-3	NG108-15	120
		C6-BU-1	121
	5HT-2	A7r5	122
	5HT-3	N1E-115	145
Glutamate	NMDA	HT-4	124
	Quisqualate	N18-RE-105	125, 126

Table 3. (contd.)

Receptor	Subtype	Cell line	Reference
Adenosine	A2	NG108-15	127, 138
	A1, A2	DDT1 MF-2	128
	A2	PC12	129, 133, 136
		PC18	129
	A2	IMR 32	130
	A1, A2	LLC PK1	131
	A1	GH3	132, 135
		1321 N1	134
		GH4 C1	137
GABA (benzodiazipine)	GABA-A	C6	140, 144
		N2a	141
	Benzodiaz.	CHO	142
	Benzodiaz.	NG108-15	138
	Benzodiaz.	MDCK	143
	GABA-A	NB2A	144

Table 4. Cell lines expressing voltage-activated ion channels

Ion channel	Cell line	References
Calcium	PC12	146–149
	F11	150, 151
	NG108-15	152, 172
	GH3	153, 154, 158
	A7r5	155
	A10	156
	3T3	157
	RINm5F	159
	MC3T3-E1	160
	NCB20	
	N1E-115	
	PCC4-Aza1-ECA2	162
Sodium	UCHCC1	161
	PC12	163–166
	PCC4-Aza1-ECA2	162
	IMR32	167
	N1E-115	168
	LA-N-5	169
	NTERA-2	170
	L-6	171

Table 5. Sources of cell lines

American Type Culture collection,
12301 Parklawn Drive,
Rockville,
MD 20825, USA
FAX 301 231 5826
Tel. 800-333-6078

European Collection of Animal Cell Cultures,
PHLS Centre,
Porton Down,
Salisbury SP4 0JG, UK
FAX (44) 0980 610315
Tel. (44) 0980 610391

Linscotts Directory
40 Glen Drive,
Mill Valley,
CA 94941, USA
Tel. 415-383-2666

The Human Genetic Mutant Cell Repository,
Coriell Institute for Medical Research,
Copewood and Davis Streets,
Camden,
NJ 08103, USA
Tel. 609.966 7377

References

1. Schwartz, J. P. (1988). *Brain Res.*, **427,** 141.
2. Yoshikawa, K. and Sabol, S. L. (1986). *Brain Res.*, **387,** 75.
3. Folkesson, R., Monstein, H. J., Geijer, T., Nilsson, K., and Terenius, L. (1988). *Brain Res.*, **427,** 147.
4. Byrd, J. C., Naranjo, J. R., and Lindberg, I. (1987). *Endocrinology*, **121,** 1299.
5. Verbeeck, M. A. and Burbach, J. P. (1990). *FEBS Lett.*, **268,** 88.
6. Dandekar, T. and Schulz, R. (1987). *Neuropeptides*, **9,** 25.
7. Gilbert, J. A., Knodel, E. L., Stenstrom, S. D., and Richelson, E. (1982). *J. Biol. Chem.*, **257,** 1274.
8. Rosen, H., Polakiewicz, R. D., Benzakine, S., and Bar-Shavit, Z. (1991). *Proc. Natl. Acad. Sci., USA*, **88,** 3705.
9. Verbeek, M. A., Draaijer, M., and Burbach, J. P. (1990). *J. Biol. Chem.*, **265,** 18087.
10. Mains, R. E. and Eipper, B. A. (1976). *J. Biol. Chem.*, **251,** 4115.
11. Li, W. I., Chen, C. L., and Chou, J. Y. (1989). *Endocrinology*, **125,** 2862.
12. White, A., Stewart, M. F., Farrell, W. E., Crosby, S. R., Lavender, P. M., Twentyman, P. R., et al. (1989). *J. Mol. Endocrinol.*, **3,** 65.
13. Dermody, W. C., Rosen, M. A., Ananthaswamy, R., Levy, A. G., Hixsin, C. V., Aldenderfer, P. H., et al. (1981). *J. Clin. Endocrinol. Metab.*, **53,** 970
14. McMurray, C. T., Devi, L., Calavetta, L., and Douglas, J. O. (1989). *Endocrinology*, **124,** 49.
15. Wood, J. N., Bevan, S. J., Coote, P. R., Dunn, P. M., Harmar, A., Hogan, P., et al. (1990). *Proc. R. Soc. Lond. (Biol.)*, **241,** 187.
16. Zaidi, M., Datta, H. K., Chambers, T. J., and MacIntyre, I. (1989). *Biochem. Biophys. Res. Commun.*, **158,** 214.
17. Zeytin, F. N., Rusk, S., and Leff, S. E. (1987). *Endocrinology*, **121,** 361.
18. Muszynski, M., Birnbaum, R. S., and Roos, B. A. (1983). *J. Biol. Chem.*, **258,** 11678.
19. Haun, R. S., Beienfield, M. C., Roos, B. A., and Dixon, J. E. (1989). *Endocrinology*, **125,** 850.

20. Schneider, B. S., Helson, L., Monahan, J. W., and Friedman, J. M. (1989). *J. Clin. Endocrinol. Metab.*, **69**, 411.
21. Odum, L. and Rehfeld, J. F. (1990). *Biochem. J.*, **271**, 31.
22. Zeytin, F. N., Rusk, S. F., and De-Lellis, R. (1988). *Endocrinology*, **122**, 1133.
23. Green, R. and Sheilds, D. (1984). *Endocrinology*, **114**, 1990.
24. Allen, J. M., Martin, J. B., and Heinrich, G. (1987). *Brain Res.*, **427**, 39.
25. O'Hare, M. M. and Schwartz, T. W. (1989). *Cancer Res.*, **49**, 7015.
26. Higuchi, H., Yang, H. Y., and Sabol, S. L. (1988). *J. Biol. Chem.*, **263**, 6288.
27. Hayashi, Y., Yanagawa, T., Yoshida, H., Azuma, M., Nishida, T., Yura, Y., and Sato, M. (1987). *J. Natl. Cancer Inst.*, **79**, 1025.
28. Maneckjee, R., Biswas, R., and Vonderhaar, B. K. (1990). *Cancer Res.*, **50**, 2234.
29. Yu, V. C., Eiger, S., Duan, D. S., Lameh, J., and Sadee, W. (1990). *J. Neurochem.*, **55**, 1390.
30. MacDermot, J. and Nirenberg, M. (1978). *FEBS Lett.*, **90**, 345.
31. Scheideler, M. A., Lockney, M. W., and Dawson, G. (1983). *J. Neurochem.*, **41**, 1261.
32. Chang, K-J. and Cuarecasas, P. (1979). *J. Biol. Chem.*, **254**, 2610.
33. Carr, D. J., DeCosta, B. R., Kim, C. H., Jacobsen, A. E., Guarcello, V., Rice, K. C., and Blalock, J. E. (1989). *J. Endocrinol.*, **122**, 161.
34. Kushner, L., Zukin, S. R., and Zukin, R. S. (1988). *Mol. Pharmacol.*, **34**, 689.
35. Yu, V. C., Richards, M. L., and Sadee, W. (1986). *J. Biol. Chem.*, **261**, 1065.
36. West, R. E., Freedman, S. B., Dawson, G., Miller, R. J., and Villereal, M. L. (1982). *Life Sci.*, **31**, 1335.
37. Zagon, I. S. and McLaughlin, P. J. (1990). *Neuroscience*, **37**, 223.
38. Zagon, I. S., Goodman, S. R., and McLaughlin, P. J. (1990). *Brain Res.*, **511**, 181.
39. Roth, K. A. and Barchas, J. D. (1986). *Cancer*, **57**, 769.
40. Lee, C. M., Kum, W., Cockram, C. S., Teoh, R., and Young, J. D. (1989). *Brain Res.*, **488**, 328.
41. Hochhaus, G., Yu, V. C., and Sadee, W. (1986). *Brain Res.*, **382**, 327.
42. McLawhon, R. W., Cermak, D., Ellory, J. C., and Dawson, G. (1983). *J. Neurochem.*, **41**, 1286.
43. Raue, F., Schneider, H. G., Zink, A., and Ziegler, R. (1987). **19**, 563.
44. Inoue, N. and Hatanaka, H. (1982). *J. Biol. Chem.*, **257**, 9238.
45. Ihara, H. and Nakanishi, S. (1990). **265**, 22441.
46. Findlay, D. M., Michelangeli, V. P., and Robinson, P. J. (1989). *Endocrinology*, **125**, 2656.
47. Upchurch, K. S., Parker, L. M., Scully, R. E., and Krane, S. M. (1986). *J. Bone Miner. Res.*, **1**, 299.
48. Moseley, J. M., Smith, P., and Martin, T. J. (1986). *J. Bone Miner. Res.*, **1**, 293.
49. Wohlwend, A., Malmstrom, K., Henke, H., Murer, H., Vassalli, J. D., and Fischer, J. A. (1985). *Biochem. Biophys. Res. Commun.*, **131**, 537.
50. Findlay, D. M., deLuise, M., Michelangeli, V. P., Ellison, M., and Martin, T. J. (1980). *Cancer Res.*, **40**, 1311.
51. Hunt, N. H., Ellison, M., Underwood, J. C., and Martin, T. J. (1977). *Br. J. Cancer*, **35**, 777.
52. Klueppelberg, U. G., Molero, X., Barrett, R. W., and Miller, L. J. (1990). *Mol. Pharmacol.*, **38**, 159.

53. Lignon, M. F., Bernad, N., and Martinez, J. (1991). *Mol. Pharmacol.*, **39**, 615.
54. Reyl-Desmars, F., LeRoux, S., Linard, C., Benkouka, F., and Lewin, M. J. (1989). *C.R. Acad. Sci.* (III), **308**, 251.
55. Reyl-Desmars, F., Le Roux, S., Linard, C., Benkouka, F., and Lewin, M. J. (1989). *J. Biol. Chem.*, **264**, 18789.
56. Reisine, T., Wang, H. L., and Guild, S. (1988). *J. Pharmacol. Exp. Ther.*, **245**, 225.
57. Reyl-Desmars, F., Laboisse, C., and Lewin, M. J. (1986). *Regul. Pept.*, **16**, 207.
58. Hierowski, M. T., Liebow, C., du-Sapin, K., and Schally, A. V. (1985). *FEBS Lett.*, **179**, 252.
59. Olasmaa, M., Pahlman, S., and Terenius, L. (1987). *Neurosci. Lett.*, **83**, 161.
60. Gordon, A. E., Kohout, T. A., and Fishman, P. H. (1990). *J. Neurochem.*, **55**, 506.
61. Lobaugh, L. A. and Blackshear, P. J. (1990). *Am. J. Physiol.*, **258**, C913.
62. Sheikh, S. P., O'Hare, M. M., Tortora, O., and Schwartz, T. W. (1989). *J. Biol. Chem.*, **264**, 6648.
63. Van Valen, F., Jurgens, H., Winkelmann, W., and Keck, E. (1989). *Cell Signal*, **1**, 435.
64. Tumer, J. T., Franklin, C. C., Bollinger, D. W., and Kim, H. D. (1990). *Am. J. Physiol.*, **258**, C266.
65. Waschek, J. A., Muller, J. M., Duan, D. S., and Sadee, W. (1989). *FEBS Lett.*, **250**, 611.
66. Laburthe, M., Augeron, C., Rouyer-Fessard, C., Roumagnac, I., Maoret, J. J., Grasset, E., and Laboisse, C. (1989). *Am. J. Physiol.*, **256**, G443.
67. Wessels-Reiker, M., Haycock, J. W., Howlett, A. C., and Strong, R. (1991). *J. Biol. Chem.*, **266**, 9347.
68. Francel, P. C., Miller, R. J., and Dawson, G. (1987). *J. Neurochem.*, **48**, 1632.
69. Snider, R. M. and Richelson, E. (1984). *J. Neurochem.*, **43**, 1749.
70. Yano, K., Higashida, H., Inoue, R., and Nozawa, Y. (1984). *J. Biol. Chem.*, **259**, 10201.
71. Becherer, P. R., Mertz, L. F., and Baenziger, N. L. (1982). *Cell*, **30**, 243.
72. Weiss, C. and Atlas, D. (1991). *Brain Res.*, **543**, 102.
73. Jackson, T. R., Hallam, T. J., Downes, C. P., and Hanley, M. R. (1987). *EMBO J.*, **6**, 49.
74. Francel, P. C., Keefer, J. F., and Dawson, G. (1989). *Mol. Pharmacol.*, **35**, 34.
75. Coffer, A., Sinnet-Smith, J., and Rozengurt, E. (1990). *FEBS Lett.*, **275**, 159.
76. Moody, T. W., Mahmoud, S., Staley, J., Cirillo, D., South, V., Felder, S., and Kris, R. (1989). *J. Mol. Neurosci.*, **1**, 235.
77. Singh, P., Draviam, E., Guo, Y. S., and Kurosky, A. (1990). *Am. J. Physiol.*, **258**, G803.
78. Logsdon, C. D., Zhang, J. C., Guthrie, J., Vigna, S., and Williams, J. A. (1987). *Biochem. Biophys. Res. Commun.*, **144**, 463.
79. Battey, J. F., Way, J. M., Corjay, M. H., Shapira, H., Kusano, K., Harkins, R., et al. (1991). *Proc. Natl. Acad. Sci. USA*, **88**, 395.
80. Feldman, R. I., Wu, J. M., Jenson, J. C., and Mann, E. (1990). *J. Biol. Chem.*, **265**, 17364.
81. Brown, K. D., Laurie, M. S., Littlewood, C. J., Blakely, D. M., and Corps, A. N. (1988). *Biochem. J.*, **252**, 227.

82. Zachary, I. and Rozengurt, E. (1987). *J. Biol. Chem.*, **262,** 3947.
83. Westendorf, J. M. and Schonbrunn, A. (1983). *J. Biol. Chem.*, **258,** 7527.
84. Siegel, H. N. and Lukas, R. J. (1988). *J. Neurochem.*, **50,** 1272.
85. Simasko, S. M., Durkin, J. A., and Weiland, G. A. (1987). *J. Neurochem.*, **49,** 253.
86. Gotti, C., Wanke, E., Fornasari, D., Cabrini, D., and Clementi, F. (1986). *Biochem. Biophys. Res. Commun.*, **137,** 1141.
87. Kato, E., Anwyl, R., Quandt, F. N., and Narahashi, T. (1983). *Neuroscience, 8,* 643.
88. Syapin, P. J., Salvaterrs, P. M., Engelhardt, J. K. (1982). *Brain Res.*, **231,** 365.
89. Hom, V, J., Baum, B, J., and Ambukar, I. S. (1991). *Biochem. Biophys. Res. Commun.*, **177,** 784.
90. Baron, B. M. and Siegel, B. W. (1989). *J. Neurochem.*, **53,** 602.
91. Baumgold, J. and Fishman, P. H. (1988). *Biochem. Biophys. Res. Commun.*, **154,** 1137.
92. DeGeorge, J. J., Morell, P., McCarthy, K. D., and Lapetina, E. G. (1986). *J. Biol. Chem.*, **261,** 3428.
93. Hughes, A. R. and Harden, T. K. (1986). *J. Pharmacol. Exp. Ther.*, **237,** 173.
94. Wojcikiewicz, R. J., Dobson, P. R., and Brown, B. L. (1984). *Biochim. Biophys. Acta,* **805,** 25.
95. Pinkas-Kramarski, R., Edelman, R., and Stein, R. (1990). *Neurosci. Lett.*, **108,** 335.
96. Edwards, A., Gillard, M., and Merler, E. (1989). *J. Recept. Res.*, **9,** 259.
97. McKinney, M., Stenstrom, S., and Richardson, E. (1984). *Mol. Pharmacol.*, **26,** 156.
98. Klinj, K., Slivka, S. R., Bell, K., and Insel, P. A. (1991). *Mol. Pharmacol.*, **39,** 407.
99. He, X. J., Wu, X. Z., Brown, A. M., Wellner, R. B., and Baum, B. J. (1989). *Gen. Pharmacol.*, **20,** 175.
100. Mauger, J. P., Sladeczek, F., and Bockaert, J. (1982). *J. Biol. Chem.*, **257,** 875.
101. Cornett, L. E. and Norris, J. S. (1982). *J. Biol. Chem.*, **257,** 694.
102. Ullrich, S. and Wollheim, C. B. (1989). *Acta Endocrinol. (Copenhagen),* **121,** 525.
103. Bresahan, M. R., Flordellis, C. S., Vassilatis, D. K., Makrides, S. C., Zannis, V. I., and Gavras, H. (1990). *Biochim. Biophys. Acta,* **1052,** 439.
104. Bylund, D. B. and Ray-Prenger, C. (1989). *J. Pharmacol. Exp. Ther.*, **251,** 640.
105. Kazmi, S. M. and Mishra, R. K. (1989). *Biochem. Biophys. Res. Commun.*, **158,** 921.
106. Murphy, T. J. and Bylund, D. B. (1988). *J. Pharmacol. Exp. Ther.*, **244,** 571.
107. Fishman, P. H., Nussbaum, E., and Duman, R. S. (1991). *J. Neurochem.*, **56,** 596.
108. Guest, S. J., Hoadcock, J. R., Watkins, D. C., and Malbon, C. C. (1990). *J. Biol. Chem.*, **265,** 5370.
109. Madtes, P., Kyritsis, A., and Chader, G. J. (1985). *J. Neurochem.*, **45,** 1836.
110. Abramson, S. N. and Molinoff, P. B. (1985). *J. Biol. Chem.*, **260,** 14580.
111. Ghahary, A. and Cheng, K. W. (1990). *Cell. Mol. Neurobiol.*, **10,** 337.
112. Howell, R. E., Albelda, S. M., Daise, M. L., and Levine, E. M. (1988). *J. Appl. Physiol.*, **65,** 1251.

113. Delavier-Klutchko, C., Hoebeke, J., and Strosberg, A. D. (1984). *FEBS Lett.*, **169**, 151.
114. Norris, J. S., Garmer, D. J., Brown, F., Popovich, K., and Cornett, L. E. (1983). *J. Recept. Res.*, **3**, 623.
115. Pittman, R. N. and Molinoff, P. B. (1983). *Mol. Pharmacol.*, **24**, 398.
116. Sheppard, J. R. (1977). *Proc. Natl. Acad. USA*, **74**, 1091.
117. Martin, T. J., Nahorski, S. R., Hunt, N. H., Dawborn, J. K., Loomes, R. S., and Underwood, C. E. (1978). *Clin. Sci. Mol. Med.*, **55**, 23.
118. Conner, D. A. and Mansour, T. E. (1990). *Mol. Pharmacol.*, **37**, 742.
119. Murphy, T. J. and Bylund, D. B. (1989). *J. Pharm. Exp. Ther.*, **249**, 535.
120. Yakel, J. L. and Jackson, M. B. (1988). *Neuron*, **1**, 615.
121. Ogura, A. and Amano, T. (1984). *Brain Res.*, **297**, 387.
122. Doyle, V. M., Creba, J. A., Ruegg, U. T., and Hoyer, D. (1986). *Naunyn Schmiedebergs Arch. Pharmacol.*, **333**, 98.
123. Berry-Kravis, E. and Dawson, G. (1983). *J. Neurochem.*, **40**, 977.
124. Morimoto, B. H. and Koshland, D. E. (1990). *Proc. Natl. Acad. Sci. USA*, **87**, 3518.
125. Murphy, T. H., Malouf, A. T., Sastre, A., Schaar, R. L., and Coyle, J. T. (1988). *Brain Res.*, **444**, 325.
126. Malouf, A. T., Schaar, R. L., and Coyle, J. T. (1984). *J. Biol. Chem.*, **259**, 12756.
127. Gubits, R. M., Wollack, J. B., Yu, H., and Liu, W. K. (1990). *Mol. Brain Res.*, **8**, 275.
128. Ramkumar, V., Barrington, W. W., Jacobson, K. A., and Stiles, G. L. (1990). *Mol. Pharmacol.*, **37**, 149.
129. Raskoski, R. and Roskoski, L. M. (1989). *J. Neurochem.*, **53**, 1934.
130. Abbracchio, M. P., Cattabeni, F., Clementi, F., and Sher, E. (1989). *Neuroscience*, **30**, 819.
131. Weinberg, J. M., Davis, J. A., Shayman, J. A., and Knight, P. R. (1989). *Am. J. Physiol.*, **256**, C967.
132. Delahunty, T. M., Cronin, M. J., and Linden, J. (1988). *Biochem. J.*, **255**, 69.
133. Williams, M., Abreu, M., Jarvis, M. F., and Noronha-Blob, L. (1987). *J. Neurochem.*, **48**, 498.
134. Hughes, A. R. and Harden, T. K. (1986). *J. Pharmacol. Exp. Ther.*, **237**, 173.
135. Cooper, D. M., Caldwell, K. K., Boyajian, C. L., Petcoff, D. W., and Schlegel, W. (1989). *Cell Signal*, **1**, 85.
136. Noronha-Blob, L., Marshall, R. P., Kinnier, W. J., and U'Prichard, D. C. (1986). *Life Sci.*, **39**, 1059.
137. Dorflinger, L. J. and Schonbrunn, A. (1985). *Endocrinology*, **117**, 2330.
138. Snell, C. R. and Snell, P. S. (1984). *Br. J. Pharmacol.*, **83**, 791.
139. Steinberg, T. H. and Silverstein, S. C. (1987). *J. Biol. Chem.*, **262**, 3118.
140. Majewska, M. D. and Chuang, D. M. (1985). *J. Pharmacol. Exp. Ther.*, **232**, 650.
141. Rohde, B. H. and Harris, R. A. (1982). *Brain Res.*, **253**, 133.
142. Riond, J., Vita, N., Le Fur, G., and Ferrara, P. (1989). *FEBS Lett.*, **245**, 238.
143. Beaumont, K., Moberly, J. B., and Fanestil, D. D. (1984). *Eur. J. Pharmacol.*, **103**, 185.
144. Baraldi, M., Guidotti, A., Schwartz, J. P., and Costa, E. (1979). *Science*, **205**, 821.

145. Peters, J. A., Malone, H. M., and Lambert, J. J. (1990). *Neurosci. Lett.*, **110**, 107.
146. Fujimoto, M. and Mihara, S. (1991). *Neurosci. Lett.*, **122**, 9.
147. Rausch, D. M., Lewis, D. L., Barker, J. L., and Eiden, L. E. (1990). *Cell Mol. Neurobiol.*, **10**, 237.
148. Janigro, D., Maccaferri, G., and Meldolesi, J. (1989). *FEBS Lett.*, **255**, 398.
149. Usowicz, M. M., Porzig, H., Becker, C., and Reuter, H. (1990). *J. Physiol. (Lond.)*, **426**, 95.
150. Boland, L. M. and Dingledine, R. (1990). *J. Physiol. (Lond.)*, **420**, 223.
151. Francel, P. C., Harris, K., Smith, M., Fishman, M. C., Dawson, G., and Miller, R. J. (1987). *J. Neurochem.*, **48**, 1624.
152. Creba, J. A. and Karobath, M. (1986). *Biochem. Biophys. Res. Commun.*, **134**, 1038.
153. Simasko, S. M., Weiland, G. A., and Oswald, R. E. (1988). *Am. J. Physiol.*, **254**, E328.
154. Shangold, G. A., Kongsamut, S., and Miller, R. J. (1985). *Life Sci.*, **36**, 2209.
155. Marks, T. N., Dubyak, G. R., and Jones, S. W. (1990). *Pflugers Arch.*, **417**, 433.
156. Kongsamut, S., Freedman, S. B., and Miller, R. J. (1985). *Biochem. Biophys. Res. Commun.*, **127**, 71.
157. Chen, C. F., Corbley, M. J., Roberts, T. M., and Hess, P. (1988). *Science*, **239**, 1024.
158. Kunze, D. L. and Ritchie, A. K. (1990). *J. Membr. Biol.*, **118**, 171.
159. Aicardi, G., Pollo, A., Sher, E., and Carbone, E. (1991). *FEBS Lett.*, **281**, 201.
160. Amagi, Y. and Kasai, S. (1989). *Jpn J. Physiol.*, **39**, 773.
161. Caviedes, R., Diaz, M. A., Compagnon, D., Liberona, J. L., Cury, M., and Jaimovich, E. (1986). *Brain Res.*, **365**, 259.
162. Kubo, Y. (1989). *J. Physiol. (Lond.)*, **409**, 497.
163. Kalman, D., Wong, B., Horvai, A. E., Cline, M. J., and O'Lague, P. H. (1990). *Neuron*, **4**, 355.
164. Reed, J. K. and England, D. (1986). *Biochem. Cell Biol.*, **64**, 1153.
165. Garber, S. S., Hoshi, T., and Aldrich, R. W. (1989). *J. Neurosci.*, **9**, 3976.
166. Rudy, B., Kirschenbaum, B., Rukenstein, A., and Green, L. A. (1987). *J. Neurosci.*, **7**, 1613.
167. Gotti, C., Sher, E., Cabrini, D., Bondiolotti, G., Wanke, E., Mancielli, E., and Clememti, F. (1987). *Differentiation*, **34**, 144.
168. Baumgold, J. and Spector, I. (1987). *J. Neurochem.*, **48**, 1264.
169. Weiss, R. E. and Sidell, N. (1991). *J. Gen. Physiol.*, **97**, 521.
170. Rendt, J., Erulkar, S., and Andrews, P. W. (1989). *Exp. Cell Res.*, **180**, 580.
171. Haimovich, B., Tanaka, J. C., and Barchi, R. L. (1986). *J. Neurochem.*, **47**, 1148.
172. Docherty, R. J., Robbins, J., and Brown, D. A. (1991). In *Cellular neurobiology: a practical approach* (ed. J. Chad and H. Wheal), pp. 75–95. IRL Press at Oxford University Press, Oxford.

A2

Suppliers of specialist items

The postal addresses of the main branch of companies cited is shown.

Amersham International plc, White Lion Road, Amersham, Bucks HP7 9LL, UK.
ATCC, 123401 Parklawn Drive, Rockville, MD 20852-1776, USA.
Axon Instruments, 1101 Chess Drive, Foster City, CA 94404, USA.
Bio-Rad Laboratories, 3300 Regatta Bvd, Richmond, CA 94804, USA.
Boehringer Mannheim GmbH, Sandhoferstrasse 116, Postfach 310120, D6800 Mannheim, Germany.
Campden Instruments, King Street, Sileby, Loughborough, Leics LF12 7LZ, UK.
Cayman Chemicals, 2280 Peters Road, Ann Arbor, MI 48103, USA.
CBS Scientific, PO Box 856, Del Mar, CA 92014, USA.
Clontech Laboratories, 4030 Fabian Way, Palo Alto, CA 94303, USA.
Contraves, Times House Station Approach, Ruislip, Middlesex, UK.
Costar, 205 Broadway, Cambridge, MA 02139, USA.
Fisons, Bishop Meadow Road, Loughborough, Leics LE11 0RG, UK.
GIBCO/BRL Life Technologies Inc., 3175 Staley Road, Grand Island, NY 14072, USA.
ICN Flow, 3300 Hyland Ave., Costa Mesa, CA 92626, USA.
Koch-Light, Rockwood Way, Haverhill, Suffolk CB9 8PB, UK.
Eastman Kodak Co., Laboratory Research Products Division, 343 State Street, Bldg 701, Rochester, NY 14652-32512, USA.
IBF, 8510 Corridor Road, Savage, MD 20763, USA.
Immunodiagnostics, Boldon Business Park, Boldon, Tyne and Wear NE35 9PD, UK.
Innovative Chemistry, PO Box 90, Marshfield MA 02050 USA.
List Medical, 501-B Vandell Way, Campbell, CA 95008-6967, USA.
Luckham, Victoria Gardens, Burgess Hill, West Sussex RH15 9QN, UK.
Millipore, PO Box 255, Bedford, MA 01730, USA.
Molecular Probes, PO Box 22010, 4849 Pitchford Avenue, Eugene, OR 97402, USA.
Narashige, 27-9 Minami Karasuyama, 4-chome Setagaya-Ku, Tokyo, Japan.
New England Nuclear, 549 Albany Street, Boston, MA 02118, USA.
Nunc, 2000 North Aurora Road, Naperville, IL 60653, USA.

Suppliers of specialist items

Pharmacia/LKB, PO Box 175, Bjorkgatan 30, 751 82 Uppsala, Sweden.
Pierce Biochemicals, PO Box 117, Rockford, IL 61105, USA.
Perkin Elmer Cetus, 761 Main Avenue, Norwalk, CT 06859, USA.
Promega Biotec, 2800 Woods Hollow Road, Madison, WI 53711-5399, USA.
Sarstedt, PO Box 468, Newton, NC 28658-0468, USA.
Schleicher & Schull, 10 Optical Avenue, Keene, NH 03431, USA.
Sigma, PO Box 14509, St Louis, MI 63178, USA.
Stratagene, 11099 North Torrey Pines Road, La Jolla, CA 92037, USA.
Sutter Instrument Co., PO Box 392, St. Raphael, CA 94912, USA.
Whatman Inc., 9 Bridwell Place Clifton, NJ 07014, USA.
Worthington Biochemical Corp., Halls Mill Rd., Freehold, NJ 07728, USA.

Index

acetyl–beta–methylcholine, and $[Ca^{2+}]_i$ 170
acetylcholinesterase
  biosynthesis in post–natal septal cell lines 21
  septal cell lines 13–15
adenosine receptors, cell lines 253
adhesion
  recovery 221–3
  see also cell adhesion molecules; intercellular adhesion
adhesion assays 218, 224–36
  recovery medium 222–3
aggregation see intercellular cell adhesion
alpha-1 and alpha-2 adrenergic receptors, cell lines 252
amino acids
  charges and relative mobilities 154
  see also phosphoamino acids
amplifiers, in electrophysiology 121
antibiotics, in culture media 82
antibodies, specificity, RIA, cyclic nucleotides 110
antisera, cyclic nucleotides, raising and testing specific antisera 111–12
arachidonic acid metabolites, detection 113
assays
  adhesion assays 218
    recovery 221–3
  beta-galactosidase assay 201–2
  Bradford assay 112, 202
  cell dissociation methods 218
  chloramphenicol acetyltransferase 28, 199, 200–1
  DNA mobility shift assay 204–6
  DNase I footprinting assay 206–9
  flow cytometry, assay by flow cytometry 174
  immunofluorescence assays 49–50, 93–4
  intercellular adhesion assays 224–36
  methylation interference 209–11
  neurite outgrowth assays 237–44
  nuclear run-on assay 198
  promoter activity 197–202
  RNase protection assay 186–90
  tumourigenicity 45
astrocytes
  cell cultures 31
  serum-free media available 70, 71
attachment factors, in serum-free media 65–6

beta-adrenergic receptors, cell lines 252
beta-galactosidase assay 201–2
blastocysts, harvesting 102
bombesin receptor, cell lines 251
Bottenstein-Sato N2 medium 30
Bradford assay, protein concentration 112, 202
bradykinin receptor, cell lines 251
bradykinin–evoked ion fluxes 116, 126
brain–derived neurotrophic factor 15
buffers
  $[Ca^{2+}]$ buffer 172–3
  cacodylate 234
  chromatography 147
  electrophoresis 147
  elution buffers 190
  hybridization 190
  immunoprecipitation 142
  lysis 136
  pulsing buffer 37

calcitonin
  calcitonin receptor 251
  cell lines 25
$[Ca^{2+}]_i$, elevation, result of receptor activation 162–3
$[Ca^{2+}]$, measurement of ion fluxes 115–17
$[Ca^{2+}]$ buffer 172–3
Ca^{2+}–dependent/independent intercellular adhesion 219–21
Ca–Mg-free Tyrode's solution 5
calcium chelation, adhesion assays 218
calcium indicator, indo-1 163–6
calcium channels, cell lines 253
calcium phosphate
  precipitation, transfection with oncogenes 35–6
  testing promoter activity 200
cAMP, cGMP see cyclic nucleotides
Carnoy fixative, karyotype analysis 45
carrier proteins, in serum-free media 66
CCK
  CCK receptor 251
  cell lines 250
CD/CI adhesive mechanisms see Ca^{2+}–dependent/independent intercellular adhesion
cDNA synthesis 191
  and PCR 191

Index

cell adhesion molecules
  anti–CAM titre 232–3
  cell dissociation methods 218–19
  measurement of specific adhesion 232–6
  myelin protein Po 234
  neural N–CAMs 218, 232–6
  regulation of expression 217, 245
cell aggregation kinetics 227
  collision efficiency coefficient [E0] 228–9
  shear rate 228
cell cultures
  astrocytes, primary and secondary cultures 31
  characterization, differentiation and functional properties 47–50
  freezing protocol 47
  loading wth indo-1 165–6
  media 30
  neurons, primary cultures 30–1
  oligodendroglial cells, enrichment 31–2
  preparation for transfection 28–32
    choice of developmental stage 29
    culture conditions 29
    transient expression vs. integration 28
  selection and growth properties 44–7
  transduction of oncogenes with recombinant retroviral vectors 38–41
  transfection with oncogenes 33–5
  transformation with oncogenic DNA viruses 32–3
cell fusion procedures *see* somatic cell fusion
cell lines
  commercial suppliers 254
  embryonal carcinoma cells 80, 85–8
  embryonal stem cells 89
  expressing neuropeptides 249–50
  hippocampal cell lines 15
  hypoxanthine ribosyltransferase–deficient neuroblastoma line 6–7
  receptor function, assay by flow cytometry 174
  septal cell lines 13–15
  STO (mouse embryonic feeder–cell line) 90–1
cell membrane, electrophysiological properties 107
centrifugation
  disruption of adhesion 225, 229
  measurement of cell adhesion 230
CGRP
  cell lines 250
  CGRP receptor 251
chimeras, production 102
chloramphenicol acetyltransferase assay 28, 199, 200–1
choline acetyltransferase, septal cell lines 11–15

chloramphenicol acetyltransferase assay 200–1
  prediction of migration of phosphopeptides 153
clonal cell lines, transduction/fusion strategy 23
collision efficiency coefficient [E0] 228–9
commercial suppliers
  cell lines 254
  specialist items 261–2
Coomassie stain technique 238–9, 242
Couette viscometry
  intercellular adhesion assays 227–9
  shear rate 228
cryostat sections, neurite outgrowth 243–4
current clamp recording 122
cyclic nucleotides
  raising and testing specific antisera 111–12
  RIA
    determination of antibody specificity 110
    measurement of cAMP and cGMP 113
    sample preparation 112
  second messengers 109

diacylglycerol
  measuring release 113–14
  second messengers 109
differentiation markers 48–50
dissociation methods 218–19
  dissociation protocol 220
  recovery after 221–3
DMSO (dimethyl sulphoxide), and *all–trans-* retinoic acid, in EC cell culture medium 84
DNA mobility shift assay 204–6
  labelling oligonucleotide probes 205–6
DNA–binding transcription factors
  identification 203–13
  methylation interference 209–11
DNase I footprinting assay 206–9
dorsal root ganglion (DRG) neurons, on 3T3 monolayers preparation 239
dot blotting, and hybridization with labelled RNA 182–3, 196–7
drug application
  electrophysiology 126–30
    bath application 128
    iontophoresis 126–8
    microperfusion, U-tube and flow pipes 129–30
    'puffer' drug application 129
Dulbecco's modified Eagle's medium 5, 6
  for EC cell lines 81
  supplements 30

Index

for SV40 stocks 33
dynorphin, cell lines 249

Edman cleavage 152
EGTA, calcium chelation 171–3, 176
eicosanoids
  measuring release 113–14
  second messengers 109
electrode pullers 120
electrophoresis
  buffers 147
  glucosephosphate isomerase 10–11
  phosphoproteins 140–1, 149–50
  prediction of migration of phosphopeptides 153
  RNase protection assay 188
electrophysiology
  drug responses, characterization 118
  neurotransmitters, neuromodulators and drugs 117–19
  principles 105–7
  recording techniques 119–30
    current voltage relationships, reversal potential 123–4
    drug application methods 126–30
    equipment 120–1
    patch–clamp recording 119, 124–6
    sharp microelectrode 119, 121–3
    second messengers, correlation with altered ion fluxes 130–1
electroporation 36–7, 99–100
  medium 99–100
  preparation of ES cells and DNA 99
ELISA techniques 240, 241–2
embryonal carcinoma cells
  cell lines 80
    1003 80, 86
    BCC7S/1009 80, 85
    P19 80, 85
    PCC3 and PCC4 80, 85–6
    TERA-2 80, 86–8
  differentiation 83–4
  growth and maintenance 80–3
  nullipotency 79
  stem cell markers 92–4
  surface antigen markers 93
embryonal stem cells
  cell lines 89
  differentiation 92
  genetic manipulation 79
  growth and maintenance 89–92
  production of mutant mice by homologous recombination 94–102
  as tools in embryology 77–9
embryonic brain cells
  serum-free media available 70
  somatic cell fusion 4–8

enkephalins, cell lines 249
epidermal growth factor, in serum-free media 64

Fab' fragments, preparation 232–3
fetal calf serum, problems 55–6
fibroblast growth factors, in serum-free media 64
fibronectin, in serum-free media 65
flow cytometry
  advantages 177–8
  hardware
    calibration 170–2
    data display and analysis 168–70
    optics and electronics 166
    sample delivery system 167–8
  preparation of [Ca^{2+}] buffer 172–3
fluorescent indicators, neurotransmitter receptors 161–6
freezing protocol, cell cultures 47

G418 solution 101
G-proteins, and second messengers 107–10
GABA, gating ion channels, membrane permeability 118
GABA receptors, cell lines 253
gel retardation assay 204
gel samples
  alkali treatment
    polyacrylamide gels 144
    PVDF membrane 144–5
  determination of radiolabelled phosphoamino acid 145–7
gene expression
  detection and quantitation of RNA 182–94
  measurement of transcription rates 194–7
  promoter activity 197–202
  regulation, neuronal cell lines 181–214
gene targeting
  design of constructs 95–7
  introducing DNA into embryonal stem cells 97–100
  isolation and characterization of clones 101
  neomycin-resistance gene (neo^r) 95–7
  positive/negative selection 95
  production of chimeras 79, 102
  selection of G418-resistant colonies 100–1
germ cell tumours 77–8
Gimmel factor, in serum-free media 64
glial cells
  serum-free media available 70, 71
  see also astrocytes; oligodendrocytes

265

Index

glial fibrillary acidic protein
   detection 505
   differentiation marker 48–50
glioma cell line, serum-free media available 70
glucose phosphate isomerase, electrophoresis 9–11
glutamate receptors, cell lines 252
Goldman equation 106
GPI isozymes *see* glucose phosphate isomerase
Gq ganglioside, detection 505
growth factors, in serum-free media 63–5
C-14 guanidine, measurement of ion fluxes 115

Ham's media 58–60
HAT medium, selection against hypoxanthine ribosyltransferase-deficient neuroblastoma line 8
hippocampal cell lines
   generation 15
   non-NGF trophic lines 15
5-HT1c and 5-HT2 *see* serotonin receptor
hypothalamus, serum-free media available 70
hypoxanthine medium *see* HAT medium
hypoxanthine ribosyltransferase-deficient neuroblastoma line, (NG18TG2) 6–7

immortalization events 2
immortalized clones, selection 44
immunofluorescence assays
   cell surface antigens 93–4
   detection of surface markers 49–50
immunoglobulins, testing for anti-CAM titre 232–3
immunoprecipitation
   buffers 142
   labelled phosphoamino acid determination 141
   protocol 143
immunopurification 141–4
indo-1, calcium indicator 163–6
inositol phosphates
   and $[Ca^{2+}]_i$, second messengers 109
   measurement of turnover 114–15
insulin, in serum–free media 61–2
intercellular adhesion
   Ca^{2+}-dependent/independent 219–21
   cell dissociation methods 218–19
   coaggregation of differentially labelled cell populations 234–6
   measurement 223
   specific adhesion 232–6

recovery of adhesiveness 221–3
   see also cell adhesion molecules; intercellular adhesion assays
intercellular adhesion assays 224–36
   aggregation in Couette flow 227–9
   aggregation on rotators 224–7
   disruption of adhesion by centrifugation 229–32
   measurements of specific adhesion 232–6
   shear rate 228
ion channels, activity *see* electrophysiology
ion fluxes
   altered, correlation with second messengers 130
   biochemical measurement 115–17
ionomycin, changes of $[Ca^{2+}]_i$ over time 171–2
iontophoresis, drug application 126–8

kainate-sensitive receptor, calcium entry 162
karyotype analysis 9, 45–6
ketanserin binding, serotonin receptor 177

laminin, in serum-free media 65–6
LETS protein *see* fibronectin
leukaemia inhibitory factor, LIF/DIA supply 89–90
lipids, in serum–free media 66
lipofection 37–8
lysates
   denatured 136
   native cell 136
   preparation for two-dimensional gel analysis 138–41
lysis buffers 136

MCh *see* acetyl–beta–methylcholine
mesencephalic cell lines, generation 16–20
mesulergine binding, serotonin receptor 177
metabolic labelling of culture cells 133–8
   phosphorus-32 incorporation 133–4
methylation interference assay, DNA-binding transcription factors 209–11
MHC antigens, EC cell culture 93
mianserin binding, serotonin receptor 177
microelectrodes
   impaling cells 122
   see also sharp microelectrode recording
microforge, in electrophysiology 120
micromanipulator, in electrophysiology 121
microperfusion methods of drug application, U-tube and flow pipes 129–30

Index

mitomycin C, treatment 43–4
MMTV, long terminal repeats 34–5
MPTP, neurotoxicity to dopaminergic neurons 19–20
muscarinic receptors, cell lines 252
muscle, neurite outgrowth, cryostat sections of muscle 243–4
myelin protein Po, as N–CAM 234

neomycin-resistance gene (neo^r) 95–7
nerve growth factor, in serum–free media 65
nerve growth factor receptor, 'trk' protooncogenes 15
nerve growth factor–related trophic factors *see* neurotrophins
neurite outgrowth 237–44
   cryostat sections of muscle 243–4
   determination 240
   principles 237
neurite outgrowth assays 237–44
   cultured cell monolayers 238–43
   DRG neurons on 3T3 monolayers 239–40
   PC 12 cells, ELISA and image analysis 241–2
neuroblastoma
   fusion partner cells, somatic cell fusion 6–7
   hypoxanthine ribosyltransferase-deficient (NG18TG2) 6–7
   serum–free media available 70
neurofilaments, differentiation marker 48–50
neuronal cell lines
   gene expression regulation 181–214
   primary neurons, serum–free media 70–1
   serum-free media 67–70
neuropeptides
   cell lines 249–50
   neuropeptide Y, cell lines 250
   neuropeptide Y receptor 251
neurotransmitter receptors
   cell lines 252–3
   fluorescent indicators 161–6
   transfected cell lines
      applications 174–8
      identification and selection of cells 174–5
      isolation and characterization 161–78
      pharmacology 177
      physiology 175–6
neurotransmitters
   gating ion channels, membrane permeability 118
   and neuromodulators, electrophysiology 117–19
neurotrophins, BDNF and NT–3 15
nicotinic receptors, cell lines 252

nigrostriatal pathway, generation of somatic fusion cells 2–4
nitric oxide, second messengers 109
NMDA receptor, calcium entry 162
Northern blot hybridization 185, 186
Northern blotting 183–5
   RNase protection 186–90
nuclear extracts, preparation 203
nuclear run-on protocol 195–6
nuclei, measurement of transcription rates 194–7

okadaic acid, inhibitor of protein phosphatases 158
oligodendrocytes, serum-free media available 70, 72
oligodendroglial cells, cell cultures 31–2
oligonucleotide probes, labelling for DNA mobility shift assay 205–6
opioid receptors, cell lines 250–1

packaging cell lines, helper proviral sequences 41–2
patch-clamp recording 124–6
   drug application methods 126–30
   four configurations 128
   gigaseal 124–6
   impaling cells 125
peptide receptors, cell lines 250–1
peptides, synthetic 153–6
permeability coefficient 106
pheochromocytoma, PC12 cells 19
phorbol 12–myristate 13–acetate, and protein kinase C 156
phosphatases, inhibition 141
phosphoamino acid, charges and relative mobilities 154
phosphoamino acid, radiolabelled analysis
   attached to PVDF membrane 147–8
   layout of sample positions 150
   determination
      gel samples 145–7
      immunoprecipitates 141
phosphopeptides *see* phosphoproteins
phosphoproteins
   analysis 138–51
      electrophoresis 140–1, 149–51
      two-dimensional gel analysis 138–41
   mapping 147–51
      two-dimensional tryptic 149–50
   prediction of migration of phosphopeptides, electrophoresis and chromatography 153
phosphorus 32
   labelling cultured cells 135, 137

Index

phosphorus (*contd.*)
  Northern blotting 184
  radiation hazards 134
phosphorylation sites
  identification 151–8
    assay of protein kinases blotted onto membranes 157–8
    manual Edman degradation 152–3
    prediction of migration of phosphopeptides 153
    protein kinases and phosphatases 156–7
    secondary proteolytic digestion 152
    synthetic peptides 153–6
phosphotyrosine-containing proteins
  detection using specific antibodies 141
  enrichment 144–5
phytohaemagglutinin–P, cell fusion 7, 8
plasma proteins, concentrations in human plasma 57
PMA *see* phorbol 12-myristate 13-acetate
poly-L-lysine 230
polyacrylamide gels, alkali treatment 144
polybrene, in retroviral vector-mediated transduction 43
polyethylene glycol, cell fusion 7, 8
polymerase chain reaction
  analysis, ES cell colonies 101
  protocol 191–2
  quantitative for mRNA 193–4
post-translational regulation, protein phosphorylation mechanism 133
promoter activity
  assay 197–202
  gene expression 197–202
  transfection 197
pro-opiomelanocortin, cell lines 249
protein kinase
  in situ assay 157–8
  strategies for identification 156–7
protein kinase C, and phorbol 12-myristate 13-acetate 156
protein phosphatases
  inhibitors 158
  strategies for identification 156–7
  see also phosphoproteins
protein phosphorylation in cell lines 133–58
  identification of phosphorylation sites 151–8
  metabolic labelling of culture cells 133–8
  phosphoprotein analysis 138–51
  see also phosphoproteins; phosphorylation sites
proteins
  base hydrolysis 144
  biosynthetic labelling 137–8
  concentration, Bradford assay 202
  incorporation of isotopes 137
  see also phosphoproteins

pulsing buffer, electroporation 37
PVDF membrane, alkali treatment 144–5

recombinant retrovirus
  production of stocks 42
  titre 42–3
  see also viral oncogene transfer
all-trans-retinoic acid, EC cell differentiation 84
retroviral vector-mediated transduction 43–4
'immortalization' events 2
RNA detection and quantitation 182–94
  dot or slot blotting 182–3, 196–7
  Northern blotting 183–5
  mRNA quantitation by PCR 191–4
RNA transcription
  radiolabelled product 188–9
  template preparation 188
RNase protection assay 186–90

Sato's media 60–1, 223
  recovery of cells after dissociation 222–3
second messengers
  action on ionic conductances 108
  changes, correlation with altered ion fluxes 130–1
  cyclic nucleotides 109
  diacyl glycerol 109
  eicosanoids 109
  inositol phosphates and $[Ca^{2+}]_i$ 109
  measurement 111–15
  modulation of intracellular levels 107–10
  nitric oxide 109
selection vectors
  pSV2 TK Neo β–globin 34
  pVV12MC9 34
septal cell lines
  acetylcholinesterase 13–15
  cholinergic features 13–15
  generation 11–15
  nerve growth factor receptor expression 14–15
septohippocampal pathway
  generation of somatic fusion cells 2–4
  hippocampal cell lines, generation 15
  septal cell lines, generation 11–15
serotonin receptor
  cell lines 252
  comparison of antagonist sensitivity in NIH-3T3 cells 176
  receptor function 174
  subtypes, pharmacology 177
serum components
  concentrations in human plasma 57
  mitogenic stimulation 58

Index

serum-free media 55–73
  commercially available media 73
  development 56–61
  Ham's media 58–60
  media supplements 61–6
  neuronal cell lines 67–70
  Sato's media 60–1
  specific formulations 67–72
  undefined low–protein serum supplements 72–3
sharp microelectrode recording 121–3
  current clamp 122
  drug application methods 126–30
  voltage clamp 123
shear rate 228
simian virus 40
  large T gene, oncogene vector 39–41
  for replication and transformation 32
slot blotting, and hybridization with labelled RNA 182–3
sodium channels, cell lines 253
somatic cell fusion
  characterization of lines 9–11
  dissection of mouse primary brain cells 4–6
  fusion process 7–8
  neuroblastoma fusion partner cells 6–7
  techniques 2–4
somatostatin
  cell lines 250
  receptor 251
South-western blotting 211–13
Southern blot analysis, ES cell colonies 101
spiperone binding, serotonin receptor 177
staphylococcal nuclease 138
staphylococcal protease 152
steroid hormones, in serum-free media 62–3
STO (mouse embryonic feeder-cell line)
  monolayers, preparation 90–1
striatal cell lines, generation 20–1
substance P
  cell lines 250
  receptor 251
substantia nigra, mesencephalic cell lines 16–20
suppliers
  cell lines 254
  specialist items 261–2
synthetic peptides 154–6

targeting construct *see* gene targeting
template preparation, RNA transcription 188–9

teratocarcinomas, as tools in embryology 77–9
trace elements, in serum–free media 66
transcription factors
  DNA-binding, identification 203–13
  Oct-2 186
transcription rates, measurement 194–7
transduction/fusion strategy, clonal cell lines 23
transfection
  by electroporation 36–7
  preparation of cell cultures 28–32
  testing promoter activity 197
transferrin, in serum-free media 62
transformation, secondary, avoiding 47
'trk' protooncogenes, nerve growth factor receptor 15
trypsin
  proteolytic digestion 152
  two-dimensional tryptic phosphopeptide mapping 149–50
tumourigenicity, assay 45
two-dimensional gel analysis
  alkali treatment of gels 144–5
  phosphoproteins 138–41
  preparation of lysates 138–41
Tyrode's solution 4–5
tyrosine kinases, synthesis of peptides 155

vasoactive intestinal peptide receptor 251
viral oncogene transfer
  generation of neural cell lines 27–53
  cell culture preparation 28–32
  characterization of cell lines 47–50
  recombinant retroviral vectors 38–44
  selection of cell lines 44–7
  summary and prospects 50–2
  transfection of cells with oncogenes 33–8
  transformation 32–3
packaging cell lines
  M.MuLV 41
  pN2 41
  pZIP–NeoSV(X)1 41
transcomplementation between vector and helper sequences 39
vitronectin 66
see also fibronectin
voltage clamp recording 123

whole-cell extracts, preparation 204
whole-cell recording, pipette-filling solution 126